408-1
1997-05

D1805756

CENTRALE LANDBOUWCATALOGUS

0000 0653 9601

# SOIL
## Ecology in Sustainable Agricultural Systems

# ADVANCES IN AGROECOLOGY

## *CLIVE EDWARDS, SERIES EDITIOR*

FyTo

# SOIL
# Ecology in Sustainable Agricultural Systems

Edited by

**Lijbert Brussaard**
**Ronald Ferrera-Cerrato**

Boca Raton   New York

im   951091

**Library of Congress Cataloging-in-Publication Data**

Soil ecology in sustainable agricultural systems / edited by Lijbert
  Brussaard and Ronald Ferrera-Cerrato.
        p.   cm.
     Proceedings of a symposium held at the 15th International Congress
  of Soil Science, Acapulco, Mexico, July 10–16, 1994.
     Includes bibliographical references and index.
     ISBN 1-56670-277-1 (alk. paper)
     1. Agricultural ecology—Congresses.   2. Soil ecology—Congresses.
  3. Sustainable agriculture—Congresses.   I. Brussaard, Lijbert
  II. Ferrera-Cerrato, Ronald.   III. International Congress of Soil
  Science (15th : 1994 : Acapulco, Mexico)
  S589.7.S637   1997
  631.4′22—dc21                                                                96-30093
                                                                                    CIP

# Preface

The development of sustainable agricultural practices depends largely on promoting the long-term fertility and productivity of soils at economically viable levels by lowering fertilizer inputs in exchange for a higher dependence on biologically acquired and recycled nutrients; reducing pesticide use while relying more on crop rotations and biocontrol agents; decreasing the frequency and intensity of soil tillage; and increasing the recycling of crop residues and animal wastes. Important objectives of these approaches are to match the supply of soil nutrients with the demands of the crops (synchronization and synlocation) and to develop soil physical properties that optimize air and water transport at levels that minimize the losses of nutrients by leaching and gas transport. This requires a basic understanding of the interplay between the plant, soil structure/texture, and soil organisms/soil organic matter.

To address these important topics we organized the symposium "Role of the Biota in Sustainable Agriculture" during the 15th International Congress of Soil Science at Acapulco, Mexico, July 10–16, 1994. This volume contains the papers contributed to that symposium.

The first six chapters focus on basic studies, some reflecting the dual nature of roots and soil organic matter as sinks and sources of carbon and nutrients, others reflecting the effects of structure-following and structure-forming soil organisms on biochemical and biophysical processes. The final paper takes a more holistic approach in tying basic knowledge together at the (agro)ecosystem level with a view on developing biological management practices that optimize soil properties for sustained agricultural use.

The chapters in this volume reflect that soil biology is making rapid progress as a quantitative science. At the same time they show considerable potential for the application of soil biological knowledge to the sound management of agro-ecosystems. The growing pressure to turn this potential into reality is a challenge for both scientists and policy makers and, indeed, for the farmers in both the industrialized and the developing countries.

**Lijbert Brussaard**
Wageningen, The Netherlands

**Ronald Ferrera-Cerrato**
Montecillo, Mexico

# Contributors

**Damian O. Asawalam**
International Institute of Tropical
  Agriculture
Ibadan, NIGERIA

**Mike H. Beare**
New Zealand Institute for Crop and
  Food Research (CASC)
Christchurch, NEW ZEALAND

**Gerard Brouwer**
DLO Research Institute for
  Agrobiology and Soil Fertility
  (AB-DLO)
Haren, THE NETHERLANDS

**Lijbert Brussaard**
Agricultural University
Department of Terrestrial Ecology
  and Nature Conservation
Wageningen, THE
  NETHERLANDS

**Claire Chenu**
INRA
Station de Science du Sol
Versailles, FRANCE

**Johannes W. Dalenberg**
DLO Research Institute for
  Agrobiology and Soil Fertility
  (AB-DLO)
Haren, THE NETHERLANDS

**Ronald Ferrera-Cerrato**
Colegio de Postgraduados
Microbiology Area
Programa de Edafologia
Instituto de Recursos Naturales
Montecillo, MEXICO

**Jan Hassink**
DLO Research Institute for
  Agrobiology and Soil Fertility
  (AB-DLO)
Haren, THE NETHERLANDS

**Stefan Hauser**
Resource and Crop
  Management Division
International Institute of
  Tropical Agriculture
Humid Forest Station
Yaoundé, CAMEROON

**Francisco J. Matus**
Escuela de Agronomia
Universidad de Talca
Talca, CHILE

**Lindsey Norgrove**
King's College
University of London
London, UNITED KINGDOM

**Jesús Pérez-Moreno**
Programa de Edafología
Instituto de Recursos Naturales
Microbiology Area
Colegio de Postgraduados
Montecillo, MEXICO

**Mike J. Swift**
Tropical Soil Biology and
  Fertility Programme
UN Complex
Nairobi, KENYA

**Bernard Vanlauwe**
International Institute of Tropical
 Agriculture
Ibadan, NIGERIA

**Meine van Noordwijk**
International Centre for Research
 in Agroforestry (ICRAF) SE Asia
Bogor, INDONESIA

# Contents

# CHAPTER 1

# Interrelationships Between Soil Structure, Soil Organisms, and Plants in Sustainable Agriculture

**L. Brussaard**

## INTRODUCTION

For reasons of sustainability of production and reduction of adverse effects on the environment, agriculture in many areas of the industrialized world strives for lower inputs of artificial fertilizers and pesticides and in some areas less soil tillage. Such agricultural systems rely more on the natural capacity of the soil to generate and maintain a "favorable" soil structure, to supply the plant with nutrients in sufficient quantities at the right time (synchronization) and the right place (synlocation), and to prevent or suppress soilborne pests and diseases. In these processes the soil biota, that is, roots and soil organisms, play an important part. The contributions of the soil biota to soil structure and soil physical properties and to the dynamics of carbon and nutrients, in particular nitrogen, were the focus of the Dutch Programme on Soil Ecology of Arable Farming Systems. In this program soil ecosystem functioning in integrated and conventional arable agriculture was compared as practiced on a silt loam soil at the Dr. H. J. Lovinkhoeve experimental farm at Marknesse in one of the polders of The Netherlands (Brussaard et al., 1988; Kooistra et al., 1989). These systems will be henceforth referred to as INT and CONV, respectively.

In this program a 4-year rotation of winter wheat, sugar beet, potatoes, and spring barley was practiced on a calcareous silt loam soil (Typic Fluvaquent with pH-KCl of 7.5; $CaCO_3$ 9%; sand 12%, silt 68%, clay 20%; average annual rainfall 740 mm). INT differed from CONV in the use of pesticides

1-56670-277-1/97/$0.00+$.50
© 1997 by CRC Press LLC

(based on observations vs. calendar; no soil fumigation vs. nematicides against potato cyst-nematodes) and fertilization (manures in addition to inorganic fertilizer and crop residues vs. inorganic fertilizer and crop residues only; nitrogen fertilizer in INT: 50 to 65% of CONV, depending on crop; C input on average in INT 2400, in CONV 1600 kg ha$^{-1}$ yr$^{-1}$). Although it is hardly practiced in The Netherlands, we included reduction of soil tillage in our design because tillage affects the soil biota probably more than agrochemicals (Doran and Werner, 1990). The soil of INT was less intensely tilled than that of CONV, viz. to 12 to 15 cm depth instead of 20 to 25 cm depth, depending on the crop. Further details on crop management are mentioned by Lebbink et al. (1994) and Van Faassen and Lebbink (1994). The INT and CONV management were each applied since 1985 on fields that had received 3270 or 1856 kg C ha$^{-1}$ yr$^{-1}$ during 32 years of previous management, resulting in organic matter contents in the topsoil of 2.8 and 2.2% and total N contents of 0.15 and 0.10% (Lebbink et al., 1994). INT was practiced on fields with the initially high organic matter and total N contents (INTA) and on fields with the initially low organic matter and total N contents (INTB). The same holds for CONV (CONVA and CONVB). Since 1987 INT was also practiced on fields with an initial organic matter content of 2.4% (Kooistra et al., 1989) with further reduction of the depth of soil tillage to 7 cm (MTnew). In some cases additional observations were made on a former grassland and on an arable farming system that had been under minimum tillage for 18 years, but had otherwise been managed as conventional (MTold). We anticipated that the 1985 high and low levels of organic matter and total N would at least be maintained in INTA and CONVB, respectively, whereas INTB and CONVA were expected to converge in organic matter and total N contents. During 6 years of observation this indeed turned out to be the case (Van Faassen and Lebbink, 1994). Most observations on soil biological and soil physical properties and processes were obtained from the fields with the initially high organic matter level, which were undergoing integrated management (INTA), and from the fields with the initially low organic matter level, which were under conventional management (CONVB).

This chapter will deal with the research objectives and some hypotheses and results, followed by practical and research implications.

## OBJECTIVES AND HYPOTHESES

Long-term objectives of the program were as follows (Brussaard et al., 1988):

1. Tuning of the nutrient supply of the soil to the nutrient demand of the plant.
2. Enhancement of the contribution of soil organisms to soil structure formation.

Against this background the following subsidiary objectives were as follows:

1. To trace the mechanisms that regulate pools and flows of carbon and nitrogen in the soil–crop ecosystem.
2. To gain an understanding of the interactions between soil organisms and soil structure.

Only the results of ad 2 are reviewed in this chapter. The results of ad 1 are reviewed in Brussard, 1994.

## RESULTS AND DISCUSSION

**Hypothesis #1a** — Porosity in general, the proportion of existing pores modified by the soil fauna, and the proportion of new pores formed by the soil fauna are higher in INTA than in CONVB because of the higher organic matter content and the higher biological activity.

**Hypothesis #1b** — As a result the soil in CONVB is more susceptible to compaction, expressed as a less stable soil structure and more horizontally oriented voids.

The higher organic matter content in INTA (2.8%) than in CONVB (2.2%) at the start of the program in 1985 was retained during the following 6 years (Van Faassen and Lebbink, 1994). The biomass and activity of soil organisms, in particular the soil fauna, likewise were higher in INTA than CONVB (e.g., Brussaard, 1994). The bulk density in the top 25 cm of soil varied between 1.2 and $1.5 \times 10^3$ kg $\cdot$ m$^{-3}$, the value in INTA being consistently $0.1 \times 10^3$ kg $\cdot$ m$^{-3}$ lower than that in CONVB (De Vos et al., 1994). In 1990 INTA and CONVB differed little in microporosity (i.e., the volume of soil, constituted by pores with diameter <30 μm), but considerably in macropores (>30 μm in diameter) (Figure 1). Macroporosity in the topsoil of INTA was much higher in 1990 than in 1987, whereas in CONVB it remained similar (Boersma and Kooistra, 1994). Analysis of soil thin sections made it possible to discriminate between the origins of voids. In 1990 both in INTA and CONVB most of the voids were due to tillage, but in INTA the percentage of voids created or modified by the soil biota was clearly higher than in CONVB (Figure 2). In INTA the impact of soil organisms on the macroporosity was visible in less than 2% of the voids in 1987, increasing to more than 5% in 1990 (Boersma and Kooistra, 1994). The percentage of macropores that was connected to the soil surface, as observed by blue-staining of pore walls after application of a methylene-blue solution on the soil surface, was not very different between INTA and CONVB (Figure 1). The development of porosity and the origin of pores in INTA is reminiscent of those in an 18-year-old minimum tillage arable farming system (MTold) that was studied for comparison on the same soil and farm in 1987 (Boersma and Kooistra, 1994).

The topsoil of INTA had a subangular, blocky structure; the basic soil structure of CONVB was also subangular and blocky, but two angular, blocky layers occurred, one below the seedbed (8 to 15 cm deep) and one below the

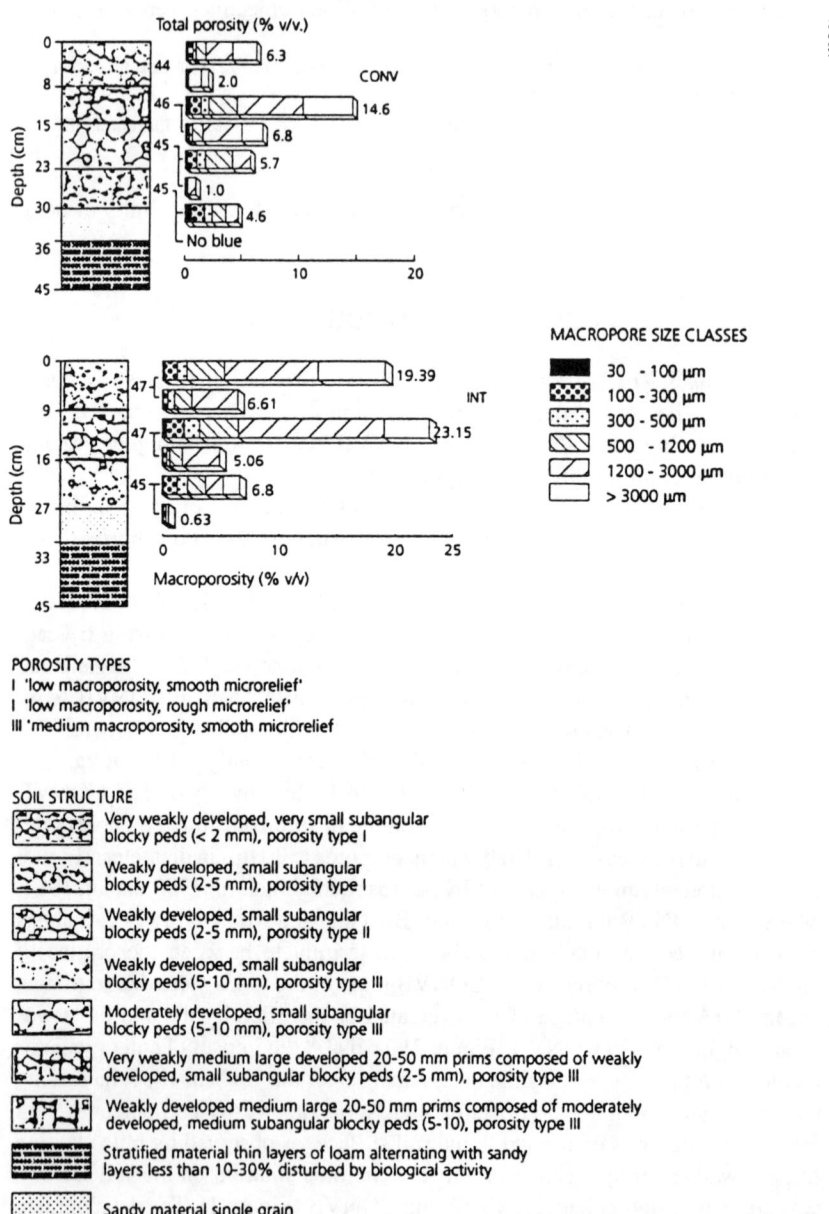

**Figure 1**   Total porosity (figures), macroporosity (bars), and macroporosity of pores connected to the surface as shown by methylene-blue staining (second bar at each depth) in INTA and CONVB in 1990. CONVB = conventional farming system since 1985 on low-organic matter soil; INTA = integrated farming system since 1985 on high-organic matter soil. (Adapted from Boersma, O. H. and Kooistra, M. J., 1994. *Agric. Ecosystems Environ.*, 51:21–42.)

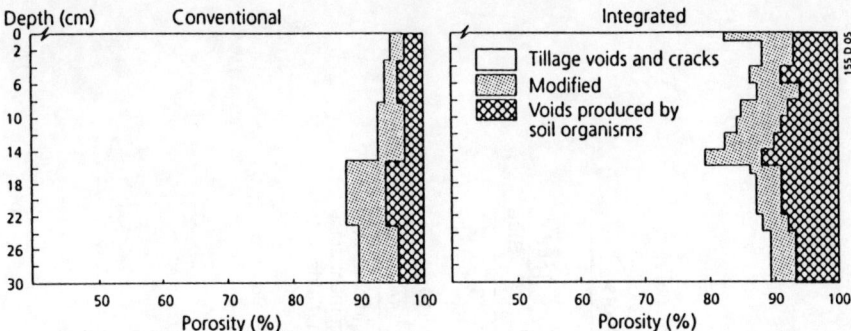

**Figure 2**    Biological impact on soil porosity in INTA and CONVB in 1990, expressed as
the percentage of macropores ($\phi$ >30 µm) of different origins. (From Boersma,
O. H. and Kooistra, M. J., 1994. *Agric. Ecosystems Environ.*, 51:21–42. ©
1994 with kind permission from Elsevier Science B.V., 1055 KV, Amsterdam,
The Netherlands.)

ploughing depth (23 to 30 cm) (Boersma and Kooistra, 1994). A detailed
analysis of pore sizes and shapes showed that the majority of voids in the 8
to 16-cm layer was accounted for by cracks in CONVB, by vughs in INTA.
Most of the cracks had a horizontal orientation in CONVB, none of them in
INTA (Schoonderbeek and Schoute, 1994). Together with the lower
macroporosity of the topsoil in CONVB, these observations indicate that the
soil was more susceptible to compaction in CONVB than in INTA. One of
the causes of the higher susceptibility of CONVB to the formation of angular,
blocky layers is its lower organic-matter content. Another one may be the
lower stability of soil aggregates (see Hypothesis #3).

**Hypothesis #2a** — Root–soil contact is more intense in CONVB than in
INTA, due to the lower porosity of the soil in CONVB than in INTA.

**Hypothesis #2b** — Root production is higher, and therefore root-derived
carbon inputs are higher in INTA than in CONVB because the crop in INTA
will explore a larger volume of soil for nutrients than the crop in CONVB and
will benefit from the higher soil macroporosity.

Using minirhizotrons, it was shown that half-way through the growing
period of 1989 the fine-root dry weight of winter wheat was significantly
higher in INTA than in CONVB in the top 20 cm and, although quantitatively
of little importance, also in the 60- to 100-cm layer. Subsequently, new root
growth was slightly higher in CONVB than in INTA (Van Noordwijk et al.,
1994). Root growth in the topsoil of CONVB may have been hampered by
the higher bulk density of the soil than in INTA. This shows up in the on-
average higher root–soil contact in thin sections of CONVB. Root–soil contact
was measured in thin sections from 15 and 25 cm depth in three classes: <1%,
1 to 99%, and >99% root–soil contact (Schoonderbeek and Schoute, 1994).
At 15 cm depth the average root–soil contact was considerably higher in
CONVB than in INTA (Table 1). In the deeper layer no difference between

Table 1   Number (N), Shape, and Percentage of Root–Soil Contact of Winter Wheat Roots

| Root-soil contact (%) | CONVB | | | | | | INTA | | | | | |
|---|---|---|---|---|---|---|---|---|---|---|---|---|
| | Round | | Elongated | | Total | | Round | | Elongated | | Total | |
| | N | Blue | N | Blue | N | Blue | N | Blue | N | Blue | N | Blue |
| <1 | — | — | — | — | — | — | 2 | 2 | 1 | — | 3 | 2 |
| 1–99 | 10 | — | 3 | — | 13 | — | 9 | 2 | 5 | 1 | 14 | 3 |
| >99 | 14 | — | 13 | — | 27 | — | 6 | — | 4 | — | 10 | — |
| Total | 24 | — | 16 | — | 40 | — | 17 | 4 | 10 | 1 | 27 | 5 |

*Note:* Blue = methylene-blue stained walls, indicating connection to the surface. Observations in horizontal thin sections taken in 1990 from 15 cm depth in fields with conventional farm management since 1985 on low–organic matter soil (CONVB) or integrated farm management on fields with high–organic matter soil (INTA).

Data from Schoonderbeek, D. and Schoute, J. F. T., 1994. *Agric. Ecosystems Environ.,* 51:89–98.

Table 2    Macroporosity, Winter Wheat Root Density and Root Orientation as Observed in Thin Sections in Topsoils of CONVB and INTA (in 1990)

| Depth (cm) | Macropores (%) | | | | Root density ($cm^{-2}$) | | Vertical/horizontal (%) | |
|---|---|---|---|---|---|---|---|---|
| | CONVB | | INTA | | CONVB | INTA | CONVB | INTA |
| | Total | Blue | Total | Blue | | | | |
| 8–15 | 14.6 | 6.8 | 23.2 | 5.1 | 6.8 | 4.7 | 60/40 | 63/37 |
| 18–26 | 4.7 | 0.8 | 7.0 | 0.6 | 6.9 | 2.7 | 83/17 | 69/31 |

Note: CONVB, conventional farm management since 1985 on low-organic matter soil; INTA, integrated farm management since 1985 on fields with high-organic matter soil; N, number; Blue, methylene-blue stained walls.

Data from Schoonderbeek, D. and Schoute, J. F. T., 1994. *Agric. Ecosystems Environ.*, 51:89–98.

INTA and CONVB was observed in the distribution among classes, but in the 1 to 99% class the average root–soil contact was higher in CONVB than INTA at both depths: 65 vs. 44% at 15 cm depth and 52 vs. 33% at 25 cm depth (Schoonderbeek and Schoute, 1994). The orientation of the roots was, on average, not very different between INTA and CONVB (Table 2), but in INTA more roots were found in vertically oriented pores connected to the surface, that is, pores with blue-stained walls (Table 1).

In balance, the year production of structural-wheat root material in CONVB in 1989 was 21% lower than that in INTA (1900 vs. 1500 kg · ha$^{-1}$), while the proportion of the total production still present at harvest was 32% lower in CONVB than in INTA (1400 vs. 950 kg · ha$^{-1}$) (Van Noordwijk et al., 1994). The observation of a higher root production in the topsoil in INTA than in CONVB was not confirmed in 1990 as a higher winter wheat root density in the topsoil of CONVB than in that of INTA in the thin sections (Table 2). It should be borne in mind, however, that observations in thin sections are from a much smaller sample of soil than those from minirhizotrons. Using $^{14}C$–$CO_2$ pulse labeling in 1990, significant differences in root production of winter wheat between CONVB and INTA were not observed either (Swinnen, 1994), the estimates being approximately the same as that for CONVB by Van Noordwijk et al. (1994). One explanation may be that the differences between CONVB and INTA originated early in the growing season. This period was included in the minirhizotron observations (starting in November of the preceding year) but not in the observations with pulse labeling, which started in early May, and neglected root decomposition for the first 3 weeks. The proportion of total root production still present at harvest was also estimated as different by the two methods: 63% in CONVB and 73% in INTA according to Van Noordwijk et al. (1994) and 57% in both CONVB and INTA according to Swinnen (1994). This may have been caused by different estimates of root dynamics based on root length (minirhizotrons) or root weight (pulse labeling).

Table 3　Mean Water Stability (%) of Aggregate Size Fractions from Fields with Different Farm Management

| Fraction classification (mm) | Mean stability (%) | | | | |
|---|---|---|---|---|---|
| | INTA | INTB | CONVB | MTnew | Former grassland |
| 4.8–8 | 19.51 | 18.08 | 6.20 | 29.32 | 39.67 |
| 1–2 | 52.82 | 48.19 | 19.64 | 41.85 | 77.36 |
| 0.3–1 | 86.70 | 91.04 | 72.16 | 79.61 | 88.80 |

Note: See text for explanation of fields (INTA, INTB, etc.).

From Marinissen, J. C. Y., 1994. Agric. Ecosystems Environ., 51:75–87. (© 1994 with kind permission from Elsevier Science B.V., 1055 KV, Amsterdam, The Netherlands.)

**Hypothesis #3** — The stability of soil aggregates is higher in INT than CONV because of the higher organic matter content and the higher contribution of earthworms to the formation of stable aggregates.

For evaluation of earthworm effects on soil aggregate stability, fields with various management practices were included in the study: in addition to INTA (integrated management on the high-organic matter soil of 1985) and CONVB (conventional management on the low-organic matter soil of 1985), a former grassland, INTB (integrated management on the low-organic matter soil of 1985), and MTnew (integrated with minimum tillage since 1987 on 2.4% organic matter soil). In these five fields the water stability of aggregates was highest in the smallest size fraction distinguished and lowest in the largest (Table 3), with both differences among size fractions and among treatments being statistically significant (Marinissen, 1994). Analysis of five adjacent size fractions of former grassland soil in the range of 0.3 to 8 mm showed that there was a gradual increase in the mean water stability from 40% in the class 4.8 to 8 mm to 89% in the class 0.3 to 1 mm. In addition, since the between-field differences among the water stabilities of the 0.3- to 1-mm fractions of the five investigated treatments were small, whereas those of the 4.8- to 8-mm fraction were highly significant, the study of water stability of aggregates was concentrated on the 4.8- to 8-mm and 0.3- to 1-mm size classes. In addition to measurements in soil from CONVB (where earthworms did not occur at the start of the program in 1985), measurements were done in soil from INTA and CONVA. Here, earthworms did occur at the start of the program in 1985 (Marinissen, 1992), but the management has been less favorable for earthworms in CONVA than INTA (e.g., by more intensive soil tillage and soil fumigation against cyst-nematodes). Aggregates of 4.8 to 8 mm, but not those of 0.3 to 1 mm, from earthworm casts were significantly more stable than similarly sized aggregates collected in the field in INTA (Marinissen, 1994). From November 1989 to November 1991, INTA and CONVB were sampled 13 times. The stability of macroaggregates was very variable over the season. The highest stability was always found in autumn. Significantly more aggregates of 4.8 to 8 mm were water stable in INTA than in CONVA, with the

Table 4  Water Stability (%) of Aggregate Size Fractions (mm) from Fields with Different Farm Management

| Date | INTA | | CONVA | | CONVB | |
|---|---|---|---|---|---|---|
|  | 4.8–8 | 0.3–1 | 4.8–8 | 0.3–1 | 4.8–8 | 0.3–1 |
| Spring '87 | 19.51 | 52.82 |  |  | 18.08 | 91.04 |
| June '88 | 36.71 | 90.39 | 34.24 | 93.60 | 13.43 | 79.56 |
| Sept '88 | 66.59 | 87.54 |  |  |  |  |
| Nov '89 | 60.99 | 84.64 | 46.79 | 78.74 |  |  |
| April '90 | 31.35 | 83.10 | 24.00 | 86.00 |  |  |
| Sept '90 | 60.95 | 84.89 | 53.08 | 78.31 |  |  |
| April '91 | 31.22 | 70.42 | 17.63 | 70.06 |  |  |
| Sept '91 | 47.55 |  | 49.29 |  |  |  |
| Nov '91 | 51.79 |  | 23.62 |  |  |  |

*Note:* For explanation of fields, see text.

From Marinissen, J. C. Y., 1994. *Agric. Ecosystems Environ.*, 51:75–87. (© 1994 with kind permission from Elsevier Science B.V., 1055 KV, Amsterdam, The Netherlands.)

covariable number of worms sampled and month of sampling also being significant (Marinissen, 1994). Whereas in spring 1988, aggregate stability in the size class 4.8 to 8 mm in INTA and CONVA was similar, it remained high in INTA, but decreased in CONVA (Table 4). No such difference developed in the water stability of 0.3- to 1-mm aggregates. Water stability of macro-aggregates in CONVB was and remained lower than in INTA or INTB. The water stability of macroaggregates in CONVA was higher than in CONVB, but less than in INTA and diminished over time. These observations indicate a significant positive effect of earthworms on the water stability of macro-aggregates, which is negatively affected by the conventional management. Analysis of data from various fields indicated that differences in manuring and in soil C content (but not in clay content) also coincided with differences in water stability of macroaggregates, but the variable contributing most to the variance was the number of earthworms (Marinissen, 1994).

**Hypothesis #4a** — The water content of the soil at all water potentials is higher in INTA than CONVB due to the higher organic matter content.

**Hypothesis #4b** — The hydraulic conductivity of the soil in INTA is higher than in CONVB due to the lower bulk density, the higher macroporosity, and the less horizontal orientation of cracks.

The water retention and hydraulic conductivity characteristics of the topsoil can be of prime importance for the risk of denitrification and for the activity and interactions of soil organisms in the water and air phases of the soil. In turn, they can be affected by roots and soil organisms through their impact on soil structure, in particular the formation of surface-connected pores. The water retention characteristics of the topsoils of CONVB and INTA, as determined in the laboratory, were clearly different: at each soil water potential the water content was higher in INTA than in CONVB (Figure 3A). The hydraulic conductivity characteristics of INTA and CONVB were similar (Figure 3B),

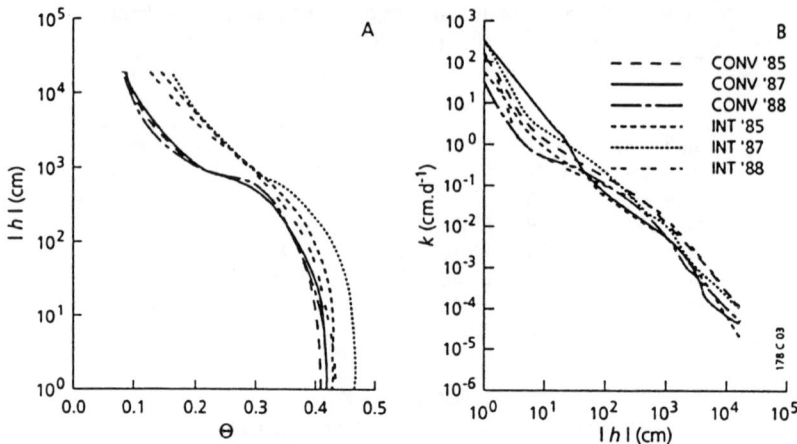

**Figure 3**   Water retention (**A**) and hydraulic conductivity (**B**) curves of the topsoils (0–25 cm) of CONVB and INTA in 1985, 1987, and 1988; Θ, water content (cm³ · cm³); *k*, hydraulic conductivity (cm · d⁻¹); I*h*I, water pressure head (cm). (From Vos, E. C. and Kooistra, M. J., 1994. *Agric. Ecosystems Environ.*, 51:227–238. © 1994 with kind permission from Elsevier Science B.V., 1055 KV, Amsterdam, The Netherlands.)

but small differences at water potentials between –100 and –2500 cm caused the calculated amount of plant-available water between field capacity and the wilting point to be lower in INTA than in CONVB (Table 5). The frequent occurrence of high groundwater levels indicated that the effective hydraulic conductivity in the subsoil was limiting the discharge of water to the drains. This means that soil physical properties of the subsoil directly influenced soil physical conditions in the topsoil, which may be important for the occurrence of denitrification (De Vos et al., 1994).

## PRACTICAL AND RESEARCH IMPLICATIONS

The results may be important for future development of farming and environmental management. Using a model that combines 30 years of weather

**Table 5**   Plant-Available Soil Water (cm³ · cm⁻³) in Topsoils of CONVB and INTA in 1985, 1987, and 1988

| *h* Ranges (cm) | Soil water fraction | | | | | |
|---|---|---|---|---|---|---|
| | CONVB '85 | CONVB '87 | CONVB '88 | INTA '85 | INTA '87 | INTA '88 |
| –100 to –2,500 | 0.232 | 0.222 | 0.234 | 0.164 | 0.195 | 0.159 |
| –100 to –16,000 | 0.283 | 0.280 | 0.286 | 0.254 | 0.268 | 0.257 |

From Vos, E. C. and Kooistra, M. J., 1994. *Agric. Ecosystems Environ.*, 51:227–238. (© 1994 with kind permission from Elsevier Science B.V., 1055 KV, Amsterdam, The Netherlands.)

data with soil water flow and crop production models, "land qualities" such as the number of workable days and the risk of anaerobic circumstances in soil (air fraction <9%) were determined. As the target, the crop that is considered most susceptible in these respects, potatoes were taken for which yield expectations were also calculated. In terms of the probability of workable days, CONVB and INTA were hardly different in critical periods of the year (i.e., spring and autumn). The risk of anaerobiosis was also similar, and the same held for the calculated yields of potato (Vos and Kooistra, 1994). Some outcomes of the model could be verified for a few years as regards anaerobiosis and yields. Gas-filled soil volumes in the topsoil of the four crops in the 4-year rotation were measured in the spring of 1988 and 1990: figures below 9% frequently were obtained, in particular in potatoes in the wet year 1988 (Figure 4). Differences between INTA and CONVB were not significant (Lebbink et al., 1994). Average tuber yields of potato in the years 1988–1991 were virtually the same: $57 \pm 7$ Mg $\cdot$ ha$^{-1}$ $\cdot$ yr$^{-1}$ in CONVB and $58 \pm 7$ Mg $\cdot$ ha$^{-1}$ $\cdot$ yr$^{-1}$ in INTA (Lebbink et al., 1994).

Although clear differences in soil structure between INTA and CONVB developed, the study period 1985 to 1990 may have been too short to reflect those in the soil physical properties and land qualities. To the extent that the soil structure of INTA is developing to that of MTold (the 18-year-old minimum tillage arable farm management), workability may become better and the risk of anaerobiosis lower in INTA (Vos and Kooistra, 1994). The occurrence of earthworms, which are absent from MTold (Marinissen, 1992), may further improve the structure of the soil and thereby reduce the risk of yield reduction. Root crops such as potatoes and sugar beet will, however, inevitably require considerable soil disturbance and high traffic loads on the fields — in sugar beet often under wet conditions. Given the adverse effects on earthworms and their low population growth potential, the earthworm fauna can only partially counterbalance these disturbances. The feasibility of integrated farming hence looks best in cropping systems with a low frequency of root crops.

Our research program has greatly benefited from experiences of and exchange with other agroecology programs that are concerned with related objectives such as the Arable Land Project in Sweden (e.g., Andrén et al., 1990); research on semi-arid grasslands and agroecosystems in the United States (e.g., Ingham et al., 1986a,b; Hunt et al., 1987); research on subtropical no-tillage agroecosystems in the United States (e.g., Beare et al., 1992); and research on dryland agriculture on different soil types in Canada (e.g., Juma, 1993). Neither these programs nor ours, has been fully successful in quantitatively integrating soil biota, organic matter, soil structure, and plant growth with the associated carbon and nutrient dynamics and water transport. Although it is not difficult to find many studies of integration of soil biological and soil physical knowledge at various spatial scales (e.g., Brussaard and Kooistra, 1993), we know of very few examples in which soil biological and

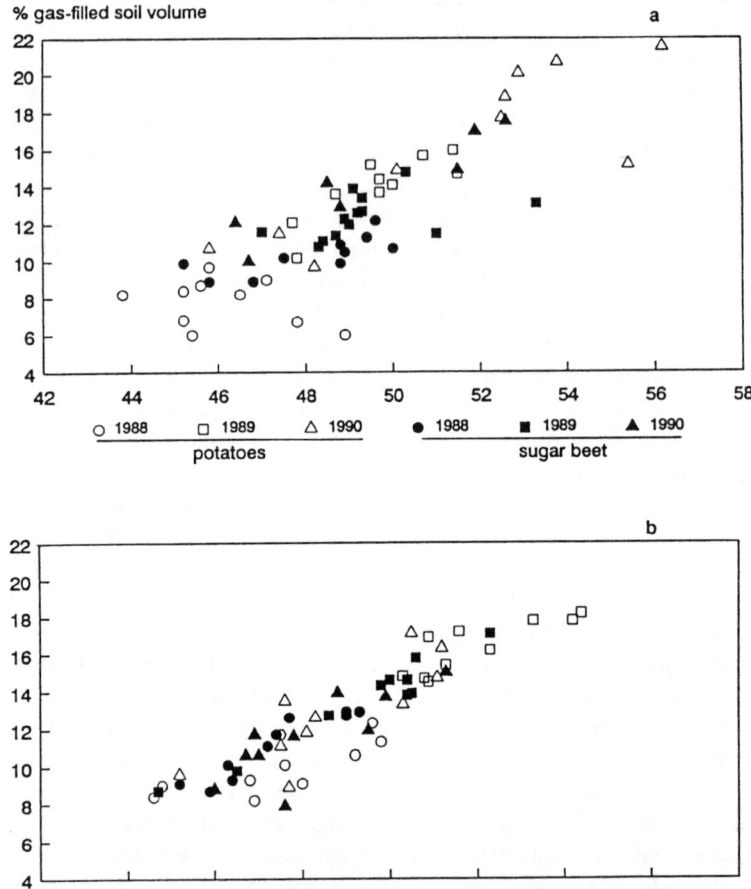

**Figure 4**    Average (n = 10) percentages of soil pore volume and gas-filled soil volume at h = −100 cm in undisturbed soil cores, taken in spring from the 12–17 and 20–25 cm layers (in potato fields the 2–7 and 12–17 cm layers below the ridge were sampled). (From Lebbink et al., 1994. *Agric. Ecosystems Environ.,* 51:7–20. © 1994 with kind permission from Elsevier Science B.V., 1055 KV, Amsterdam, The Netherlands.)

soil physical knowledge has been quantitatively integrated in descriptions of ecosystem functioning beyond conceptual models such as those of Gibbs and Reid (1988), Verberne et al. (1990), and Brussaard and Kooistra (1993). This is a first priority area for future research.

## ACKNOWLEDGMENTS

Thanks are due to M. J. Kooistra, P. A. C. Raats, and P. de Willigen for constructive comments on an earlier draft of this chapter. The research reported on in this chapter was carried out while the author was working at the DLO-Institute for Soil Fertility Research, Haren, The Netherlands, and at the Department of Soil Science and Geology of Wageningen Agricultural University, The Netherlands. This work was supported by The Netherlands Integrated Soil Research Programme and is communication no. 75 of the Dutch Programme on Soil Ecology of Arable Farming Systems.

## REFERENCES

Andrén, O., Lindberg, T., Paustian, K., and Rosswall, T., 1990. Ecology of arable land — organisms, carbon and nutrient cycling. *Ecol. Bull.,* 40:1–222.

Beare, M. H., Parmelee, R. W., Hendrix, P. F., and Cheng, W., 1992. Microbial and faunal interactions and effects on litter nitrogen and decomposition in agroecosystems. *Ecol. Monogr.,* 62:569–591.

Boersma, O. H. and Kooistra, M. J., 1994. Differences in soil structure of silt loam Typic Fluvaquents under various agricultural management practices. *Agric. Ecosystems Environ.,* 51:21–42.

Brussaard, L., 1994. The ecology of soil organisms in reduced input farming, in *Soil Biota Management in Sustainable Farming Systems,* Pankhurst, C. A., Ed., Proc. Int. Workshop on Management of the Soil Biota in Sustainable Farming Systems, Adelaide, Australia, March 15–18, 1994, 197–203.

Brussaard, L. and Kooistra, M. J., Eds., 1993. Editors' introduction, in "Soil Structure/Soil Biota Interrelationships," *Geoderma,* 56:v–vii.

Brussaard, L., Van Veen, J. A., Kooistra, M. J., and Lebbink, G., 1988. The Dutch programme on soil ecology of arable farming systems. I. Objectives, approach and some preliminary results. *Ecol. Bull.,* 39:35–40.

De Vos, J. A., Raats, P. A. C., and Vos, E. C., 1994. Macroscopic soil physical processes considered within anf agronomical and a soil biological context. *Agric. Ecosystems Environ.,* 51:43–73.

Doran, J. W. and Werner, M. R., 1990. Management and soil biology, in *Sustainable Agriculture in Temperate Zones,* Francis, C. A., Flora, C. B., and King, L. D., Eds., John Wiley & Sons, New York, 205–229.

Gibbs, R. J. and Reid, J. B., 1988. A conceptual model of changes in soil structure under different cropping systems. *Adv. Soil Sci.,* 8:123–149.

Hunt, H. W., Coleman, D. C., Ingham, E. R., Ingham, R. E., Elliott, E. T., Moore, J. C., Rose, S. L., Reid, C. C. P., and Morley, C. R., 1987. The detrital food web in a shortgrass prairie. *Biol. Fertil. Soils,* 3:57–68.

Ingham, E. R., Trofymow, J. A., Ames, R. N., Hunt, H. W., Morley, C. R., Moore, J. C., and Coleman, D. C., 1986a. Trophic interactions and nitrogen cycling in a semi-arid grassland soil. I. Seasonal dynamics of the natural populations, their interactions and effects on nitrogen cycling. *J. Appl. Ecol.,* 23:597–614.

Ingham, E. R., Trofymow, J. A., Ames, R. N., Hunt, H. W., Morley, C. R., Moore, J. C., and Coleman, D. C., 1986b. Trophic interactions and nitrogen cycling in a semi-arid grassland soil. II. System responses to removal of different groups of soil microbes and fauna. *J. Appl. Ecol.,* 23:615–630.

Juma, N. G., 1993. Interrelationships between soil structure/texture, soil biota/soil organic matter and crop production. *Geoderma,* 57:3–30.

Kooistra, M. J., Lebbink, G., and Brussaard, L., 1989. The Dutch programme on soil ecology of arable farming systems. II. Geogenesis, agricultural history, field site characteristics and present farming systems at the Lovinkhoeve experimental farm. *Agric. Ecosystems Environ.,* 27:361–387.

Lebbink, G., Van Faassen, H. G., Van Ouwerkerk, C., and Brussaard, L., 1994. The Dutch programme on soil ecology of arable farming systems: farm management, monitoring programme and general results. *Agric. Ecosystems Environ.,* 51:7–20.

Marinissen, J. C. Y., 1992. Colonization of new habitats by earthworms. *Oecologia,* 91:371–376.

Marinissen, J. C. Y., 1994. Earthworms populations and stability of soil structure in a silt loam soil of a recently reclaimed polder in The Netherlands. *Agric. Ecosystems Environ.,* 51:75–87.

Schoonderbeek, D. and Schoute, J. F. T., 1994. Root and root-soil contact of winter wheat in relation to soil macroporosity. *Agric. Ecosystems Environ.,* 51:89–98.

Swinnen, J., 1994. Rhizodeposition and turnover of root-derived organic material in barley and wheat under conventional and integrated management. *Agric. Ecosystems Environ.,* 51:115–128.

Van Faassen, H. G. and Lebbink, G., 1994. Organic matter and nitrogen dynamics in conventional versus integrated rable farming. *Agric. Ecosystems Environ.,* 51:209–226.

Van Noordwijk, M., Brouwer, G., Koning, H., Meijboom, F. W., and Grzebisz, W., 1994. Production and decay of structural root material of winter wheat and sugar beet in conventional and integrated cropping systems. *Agric. Ecosystems Environ.,* 51:99–113.

Verberne, E. L. J., Hassink, J., De Willigen, P., Groot, J. R. R., and Van Veen, J. A., 1990. Modelling organic matter dynamics in different soils. *Neth. J. Agric. Sci.,* 38:221–228.

Vos, E. C. and Kooistra, M. J., 1994. The effect of soil structure differences in a silt loam soil under various farm management systems on soil physical properties and simulated land qualities. *Agric. Ecosystems Environ.,* 51:227–238.

# CHAPTER 2

# Interactions Between Soil Biota, Soil Organic Matter, and Soil Structure

J. Hassink, F. J. Matus, C. Chenu, and J. W. Dalenberg

## INTRODUCTION

Physical protection mechanisms are important determinants of the stability of organic matter in soils. This is partly based on the observation that the turnover rates of easily decomposable compounds are much higher in liquid microbial cultures than in soils (Van Veen and Paul, 1981), and on electron microscopy studies that have demonstrated sites of accumulation of organic residues of clearly cellular origin (Foster, 1988). The phenomenon that drying of soil samples and disruption of soil aggregates can increase C and N mineralization (Richter et al., 1982; Gregorich et al., 1989) is indirect evidence of the existence of physical protection in soil. It has been concluded that there is more physical protection in fine-textured soils than in coarse-textured soils, leading to higher carbon (C) contents and larger amounts of microbial biomass (Jenkinson, 1988; Amato and Ladd, 1992; Hassink, 1994).

Several mechanisms have been suggested to explain the physical protection of organic matter in soil against decomposition. Tisdall and Oades (1982) have mentioned the adsorption of organics on or coating by clay particles. Another explanation is the entrapment of organics in small pores in or between aggregates, which renders them inaccessible to the decomposing microbial community (Elliott and Coleman, 1988).

Soil structure may also control decomposition processes by its effect on the grazing intensity of the soil fauna on microbes. The soil fauna may stimulate microbial growth rates through grazing (Clarholm, 1985; Coleman et al., 1978), and predation of microbes by protozoa and nematodes has been suggested as an important mechanism of nutrient turnover in soil (Coleman et al., 1978; Elliott et al., 1980). A large proportion of bacteria may occupy pores <3 μm (Kilbertus, 1980), while protozoa and nematodes are restricted to larger pores. This means that a large part of the bacterial population will be physically separated from the protozoa and nematodes in soil (Postma and Van Veen, 1990).

To understand how structure can affect soil organic matter turnover it is important to know how primary soil particles and organic matter interact. Tisdall and Oades (1982) present a soil structure model describing the association of organic matter with free primary particles (i.e., sand, silt, and clay), microaggregates, and macroaggregates. The basic structural units in soils are considered to be microaggregates (clay-polyvalent, metal–organic matter complexes, <250 μm in diameter) that are water stable and not affected by agricultural practices (Edwards and Bremner, 1967; Tisdall and Oades, 1982). These microaggregates are linked together to form macroaggregates, which are defined as being larger than 250 μm. Macro- and microaggregates contain primary particles, organics, and pores of different size.

Organic matter turnover and microbial activity have been studied in different soils. It is generally found that the turnover of organic matter and microbial activity is less in fine-textured soils than in coarse-textured soils (Ladd et al., 1993; Juma, 1993). In some studies, however, no effect of soil texture on C turnover and the microbial activity was found (Gregorich et al., 1991; Hassink et al., 1993a).

To improve the understanding of the interactions between soil biota, soil organic matter, and soil structure and to explain these contradictory results, it is necessary to integrate the different concepts of physical protection. The aim of the present chapter is to study, integrate, and quantify the importance of the different mechanisms of physical protection for soils of different texture.

## MATERIALS AND METHODS

### Soils

Pore size distribution, contents of primary particles (<2 μm, 2–20 μm, and >50 μm; clay, silt, and sand, respectively) and aggregate classes (microaggregates <20 μm, 20–50 μm, and 50–250 μm, and macroaggregates >250 μm), biomass of microflora and microfauna, size fractions of organic matter, and C mineralization of the soils were studied. Information on the exact location of pores, organics, and microbes was obtained using scanning electron microscopy (SEM).

Most observations were made on the top 10 cm of sandy, loamy, and clayey grassland soils and on arable sandy and clay soils. The grassland soils had been under grass for at least 8 years. They were rotationally grazed by dairy cattle and received 400 to 500 kg fertilizer-N ha$^{-1}$ yr$^{-1}$. The arable soils had been under a rotation of cereals, sugar beet, and beans for the last 8 years. Characteristics of these soils are given in Tables 1, 2, and 3. The results of

**Table 1  Selected Physical Properties of the Different Grassland and Arable Soils**

| Soil type | Granular composition (µm) % particles | | | Composition of aggregates (µm) % of total soil | | | |
|---|---|---|---|---|---|---|---|
| | <2 | <20 | <50 | <20 | 20–50 | 50–250 | >250 |
| Grassland | | | | | | | |
| Sand | 2.4 | 4.9 | 23.5 | 8.8 | 8.6 | 58.0 | 24.6 |
| Loam | 24.1 | 36.8 | 71.7 | 31.9 | 12.2 | 19.8 | 36.1 |
| Clay | 45.8 | 66.3 | 85.1 | 39.3 | 5.2 | 5.7 | 49.8 |
| Arable | | | | | | | |
| Sand | 4.7 | 8.4 | 34.6 | 12.9 | 9.6 | 64.1 | 13.4 |
| Clay | 48.1 | 77.5 | 94.0 | 23.4 | 5.2 | 11.5 | 59.9 |

**Table 2  C Contents (%) of Different Grassland and Arable Soils and the C Content of Their Aggregates and Primary Particles**

| Soil type | Total soil | Aggregate classes (µm) | | | Primary particles (µm) | | |
|---|---|---|---|---|---|---|---|
| | | 20–50 | 50–250 | >250 | <2 | <20 | >50 |
| Grassland | | | | | | | |
| Sand | 3.5 | 9.5 | 3.1 | 5.4 | 15.0 | 14.7 | 2.1 |
| Loam | 5.5 | 3.2 | 3.7 | 4.9 | 7.0 | 6.9 | 12.3 |
| Clay | 3.3 | 3.8 | 5.8 | 3.1 | 3.7 | 4.0 | 9.5 |
| Arable | | | | | | | |
| Sand | 1.6 | 1.6 | 0.6 | 1.1 | 12.4 | 9.2 | 0.7 |
| Clay | 1.8 | 1.4 | 2.0 | 1.8 | 2.1 | 1.6 | 1.8 |

**Table 3  Amount of Microbial Biomass-C in the Different Grassland and Arable Soils (mg kg$^{-1}$ Soil) and in the Clay + Silt Fraction (<20 µm; mg kg$^{-1}$ Soil Fraction)**

| Soil type | Total soil | Primary particles <20 µm |
|---|---|---|
| Grassland | | |
| Sand | 263 | 514 |
| Loam | 592 | 472 |
| Clay | 369 | 218 |
| Arable | | |
| Sand | 98 | 309 |
| Clay | 200 | 295 |

these soils have been integrated with the results of other Dutch grassland soils that have been described before (Hassink et al., 1993a,b), and with the results of other studies performed in other countries.

## Methodology

The following measurements were performed:

1. C mineralization was measured by incubating soil samples (50 g) in 1.5-liter airtight jars containing a vial of 10 ml 0.5 M NaOH. At the sampling dates, the trapped $CO_2$ was measured after precipitating the carbonate with excess $BaCl_2$.
2. Pore size distribution was calculated from the relation between soil water potential and moisture content according to Klute (1986) in undisturbed soil samples from the 2.5 to 7.5-cm layer. The effective pore neck diameter (d) was estimated from the water retention curve as: $d = 2r = -30 \times 10^{-6} \, h^{-1}$ (m).
3. C associated to primary soil particles was determined after ultrasonic dispersion of rewetted soil samples for 15 minutes with a probe-type ultrasound generating unit (Soniprep 150).
4. The biomass of bacteria was determined from counts in soil smears stained with europium-chelate; the biomass of fungi was estimated by measurements of hyphal length in agar films made by mixing 1 ml of a 1:10 diluted soil suspension and 6 ml of agar, stained with fluorescent brightener; the biomass of protozoa was estimated by the most probable number method, and the biomass of nematodes was counted after isolation by elutration.
5. Microbial biomass C was determined by the chloroform fumigation extraction method (Vance et al.,1987).
6. The amount of water-stable aggregates was determined by wet-sieving 100 g of field-moist samples on 250, 50, and 20 μm sieves that were placed on a vibrator machine for 4 minutes under a continuous water flow of 1.1 liter minute$^{-1}$.

The first four measurements have been described by Hassink et al. (1993a), and the last measurement by Matus (1997).

SEM was performed with low-temperature techniques in soils and aggregates previously incubated at –10 kPa. The procedure has been described by Chenu (1993).

When the amount and the activity of the microbial biomass were determined in primary soil particles, field-moist samples were not sonicated, but wet-sieved, crushed by hand, and pushed through a sieve with a mesh size of 150 μm until the water passing the sieve became clear. It was assumed that this had less effect on the microbial biomass than sonication. A second sieve with a mesh size of 20 μm was installed under the top sieve of 150 μm. The fractions on both sieves were collected separately. Material passing through the 20-μm sieve was collected in a bucket and concentrated afterwards. All fractions were incubated at –10 kPa.

Table 4    Mineralization Rates of C (During 14 Days at 20°C) in Dried/Remoistened Undisturbed Samples Expressed as Percentage C Mineralized per Day (×100) and Relative Increase in C Mineralization after Drying, Fine-Sieving, and Rewetting of Soil Samples in Comparison with Drying/Rewetting Only (%) in Sandy, Loamy, and Clay Grassland Soils and Some Soil Characteristics (These Soils are Different from the Soils of Table 1)

| Soil type | C mineralization | Increase in C mineralization (%) | C (%) | Granular composition % particles < | | |
|---|---|---|---|---|---|---|
| | | | | 2 μm | 20 μm | 50 μm |
| Sand | | | | | | |
| 1 | 5.6 | 12 | 4.4 | 2.6 | 4.4 | 23.5 |
| 2 | 3.5 | 16 | 4.7 | 6.5 | 9.6 | 31.5 |
| Loam | | | | | | |
| 1 | 7.1 | 23 | 3.5 | 28.2 | 45.6 | 71.2 |
| 2 | 4.1 | 85 | 5.3 | 26.2 | 36.5 | 71.7 |
| Clay | | | | | | |
| 1 | 6.6 | 230 | 6.1 | 53.5 | 76.0 | 86.4 |
| 2 | 4.7 | 185 | 5.6 | 54.6 | 77.2 | 88.5 |

To get an impression of the amount of C that was physically protected in small pores, C mineralization was determined in samples that were dried and rewetted only and in samples that were dried, finely-sieved (0.001-m mesh-size screen), and rewetted. Samples were rewetted to the original moisture content by applying a soil suspension (5 ml per 50 g soil of a 1/10 dilution) and distilled water. C mineralization was determined after incubating 50 g of the undisturbed and crushed soil samples at 20°C for 14 days. It was assumed that crushing releases some of the organic C that is physically protected in small pores (Hassink, 1992). Some characteristics of the soils are presented in Table 4.

To study the effect of soil structure and physical protection on the decomposition rate of fresh substrate, $^{14}$C-labeled glucose was applied in solution to intact and ground macroaggregates of the sandy and clay grassland soil before determining the rate of $^{14}CO_2$ production. It was observed that a part of the applied glucose penetrated into the aggregate of both the sandy and clay soil.

## RESULTS AND DISCUSSION

The results are presented in four sections. Section 1 deals with differences between the soil structure of fine- and coarse-textured soils. In the second section the dominant mechanism of physical protection of organic matter and microbes in fine- and coarse-textured soils is evaluated. In the third section we test whether the activity of the bacteria and microbial biomass is affected by soil structure. Differences in decomposition rates of fresh residues applied to fine- and coarse-textured soils are discussed in Section 4.

## Section 1. Soil Structure of Fine- and Coarse-Textured Soils

It has been stated that soil structure is the dominant control over microbially mediated decomposition processes in terrestrial ecosystems (Kuikman et al., 1990). To understand the structural restraints on decomposition processes it is necessary to characterize differences in soil structure between different soil types.

In the sandy grassland and arable soils, microaggregates with diameters between 50 and 250 μm were dominant (Table 1). In the loamy grassland soil and the clay arable and grassland soils, macroaggregates (diameters >250 μm) made up approximately 50% of total soil weight. The aggregate distribution of the loamy and clay soils was very similar to the results of Elliott (1986) when he used a similar pretreatment. The aggregate distribution of the sandy soils was quite similar to the results Christensen (1986) obtained for a sandy soil. The aggregate distribution of fine-textured soils is affected by the pretreatment of the soil, whereas for sandy soils the effect of pretreatment is small (Beare and Bruce, 1993). With increasing disruption the amount of macroaggregates decreases considerably (Elliott, 1986). Since the effect of pretreatment is different for different soils, and since various pretreatments were used in different studies, it is difficult to draw general conclusions from aggregate fractionation studies.

Pore-size distribution was determined in six grassland soils. The total pore space was lowest in the sandy grassland soils and highest in the clay grassland soils. In the loamy and clay soils, pores with diameters <0.2 μm were predominant. Also, pores with diameters between 0.2 and 1.2 μm and between 1.2 and 6 μm were more abundant in these soils than in sandy soils. In sandy soils pores with diameters between 6 and 30 μm were most abundant; the pore size class between 30 and 90 μm also made up a larger part of soil volume in the sandy soils than in the loamy and clay soils (Table 5).

SEM observations confirmed that the fabric of the clay soils was very dense (Figure 1). The pores were smaller than 10 μm, and porosity appeared

**Table 5  Pore Size Distribution (% of Soil Volume) of Some Grassland Soils (Soils are Different from Those of Table 1)**

| Soil type | \<0.2 | 0.2–1.2 | 1.2–6 | 6–30 | 30–90 | >90 | Total |
|---|---|---|---|---|---|---|---|
| | \<0.2 | 0.2–1.2 | 1.2–6 | 6–30 | 30–90 | >90 | Total |
| Sand | | | | | | | |
| 1 | 6.0 | 2.1 | 8.3 | 18.8 | 5.4 | 4.9 | 42.2 |
| 2 | 10.5 | 5.1 | 10.1 | 17.0 | 3.7 | 2.9 | 49.3 |
| Loam | | | | | | | |
| 1 | 18.4 | 8.3 | 13.4 | 0.0 | 3.0 | 5.1 | 48.1 |
| 2 | 20.1 | 11.4 | 16.1 | 0.0 | 2.6 | 8.3 | 58.7 |
| Clay | | | | | | | |
| 1 | 29.2 | 10.9 | 12.9 | 1.9 | 0.0 | 7.2 | 62.0 |
| 2 | 30.7 | 10.3 | 13.3 | 0.0 | 0.7 | 6.8 | 61.9 |

*Pore size class (μm)*

**Figure 1**  SEM observation of the internal fabric of a macroaggregate of a clayey grassland soil. White bar is 10 µm.

poorly interconnected. The fabric of the sandy soil was very different; pore sizes ranged from 1 to 150 µm and were interconnected (Figure 2).

## Section 2. Mechanisms of Physical Protection in Fine- and Coarse-Textured Soils

As it is generally found that with the same long-term input of organic material, fine-textured soils contain more organic C than coarse-textured soils (Jenkinson, 1988; Kortleven, 1963; Spain, 1990; Gregorich et al., 1991), it has been hypothesized that there is more physical protection in fine-textured soils than in coarse-textured soils. This agrees also with the postulation of Van Veen et al. (1984) that soils have characteristic capacities to preserve microorganisms and that this preservation capacity is higher in clay soils than in sandy soils. It is generally found that the amount of microbial biomass is higher in fine-textured soils than in coarse-textured soils (Ladd et al., 1985; Gregorich et al., 1991).

Hassink et al. (1993b) hypothesized that the dominant mechanism of physical protection differed for fine- and coarse-textured soils: in fine-textured soils protection of organic material by its location in small pores would be the main mechanism, whereas in coarse-textured soils organic material would be protected by its association with clay particles. This hypothesis was tested by comparing (1) the relative increase in C mineralization after destroying the structure of fine- and coarse-textured soils (assuming that this would release

**Figure 2**   SEM observation of the internal fabric of a macroaggregate of a sandy grassland soil. White bar is 100 μm.

some of the active organic matter protected in small pores), (2) by comparing the C content and the amount of microbial biomass of the clay or clay + silt fraction (<2 μm; <20 μm, respectively), and (3) by correlating the C content and C mineralization rate of different aggregate classes with their clay content.

### Effect of Disrupting the Soil Structure on C Mineralization in Fine- and Coarse-Textured Soils

C mineralization rates were determined during a period of 14 days in undisturbed and finely-sieved samples that were dried and rewetted. It was assumed that fine-sieving of soil samples would release some of the physically protected organic matter that was present in small pores and that the increase in C mineralization would give an indication of the amount of C protected in these small pores.

The increase in C mineralization decreased in the order clay, loam, and sand (Table 4). This is in agreement with the results of earlier experiments (Hassink, 1992) and confirms the hypothesis that the amount of physically protected organic C in pores was greater in fine-textured soils than in coarse-textured soils.

According to Bottner (1985) one-third to one-quarter of the biomass is destroyed after drying; after remoistening, however, the biomass is restored to the same level as before drying. In earlier experiments with the same grassland soils it was found that the amount of microbial biomass 14 days after rewetting finely-sieved samples did not differ from the initial amount of microbial biomass (Hassink, 1992), showing that there was no net contribution of the microbial biomass to C mineralization.

## Organic C and Microbial Biomass Associated with Clay Particles and/or in Aggregates

The clay particles in the sandy soils had a much higher C content than the clay particles in the loams and clays (Table 2). When the results of other Dutch grassland soils and of other studies were included, we observed a close relationship between the C content of the clay fraction in a soil and its clay content (Figure 3). SEM observations showed that clay particles were present as individual particles in coarse-textured soils, while in fine-textured soils the clay particles were coagulated. This might be the cause of the difference in C content, as in fine-textured soils less C could be adsorbed per unit of weight than in coarse-textured soils.

The C content of the clay fraction appeared to be very similar for grassland soils and arable soils. Recently reclaimed polder soils contained clay particles with lower C contents (Figure 3). The enrichment factor was obtained by dividing the C content of the clay fraction with the C content of the total soil. As also was found by Christensen (1992) and Elustondo et al. (1990), we observed that the enrichment factor decreased with clay content of the soil (Figure 4). In the sandy soils, the C content of the clay fraction was 5 to 17 times higher than that of the total soil, in the loamy soils 1 to 2 times higher, and in the clay soils it was often lower. The enrichment factor was generally higher in arable soils than in grassland soils. The lowest enrichment factors

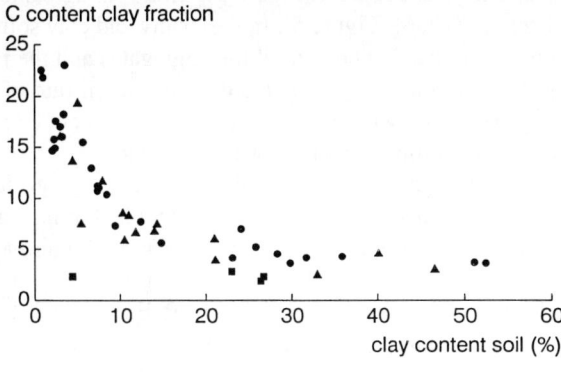

• grassland   ■ polder soils   ▲ arable

**Figure 3**   Relationship between the clay content (%) of different soils and the C content (%) of their clay fraction. (Based on data of Christensen, 1985; Elustondo et al., 1990; and results of Dutch grassland soils.)

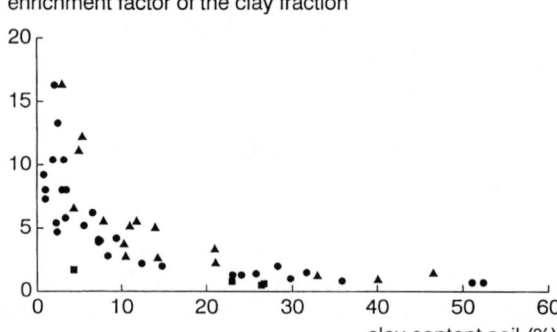

**Figure 4**　Relationship between the clay content of different soils and the C enrichment factor (% in clay fraction to % in whole soil). (Based on data of Christensen, 1985; Elustondo et al., 1990; and results of Dutch grassland soils.)

were found in recently reclaimed polder soils. The C content of the fraction <20 μm was generally slightly lower than that of the clay fraction.

The amount of microbial biomass C was higher in the loamy and clay soils than in the sandy soils (Table 3). As for total C, the clay and silt fractions in the sandy soils were enriched with microbial biomass C in comparison with the total soil, while this was not the case for the loamy and clay soils (Table 3). This higher enrichment of the clay and silt fraction with microbial biomass in coarse-textured soils than in fine-textured soils has also been observed by Jocteur Monrozier et al. (1991).

In the sandy soils the C contents of the aggregates were closely correlated with their clay or clay + silt content (Figure 5). This is in agreement with the results of Christensen (1986) (Figure 5). In the loamy and clay soils there was no correlation between the clay content of the aggregates and their C content.

We observed that in the sandy soils the decomposition rate of aggregate-C was negatively correlated with its clay or clay + silt content (Figure 6). For the loamy and clay soils this correlation was not found.

Segregation of soil organic C in micropores and microaggregates has been suggested as a major stabilizing mechanism of microbial biomass and organic matter in soil (Van Veen et al., 1985; Elliott, 1986; Kuikman et al., 1990). Christensen (1987) stated, however, that this kind of protection was not important. The conclusions of Van Veen et al. (1985) and Elliott (1986) were based on experiments with a loamy sand and a sandy loam. So the conclusions of these authors are consistent with our findings: physical protection in small pores and in aggregates is an important mechanism in fine-textured soils, but not in coarse-textured soils. In coarse-textured soils organic C and microbes are protected by their association with clay + silt particles.

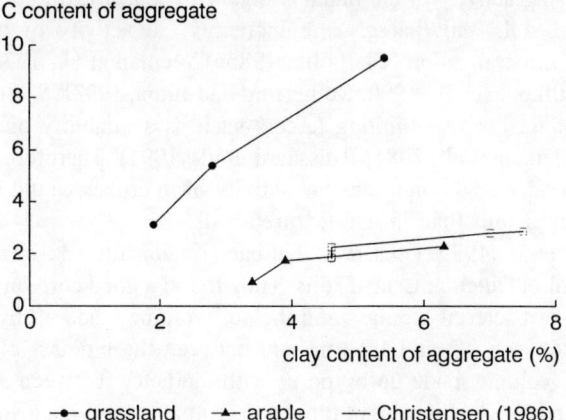

**Figure 5**  Relationship between the C content of aggregates of different classes in sandy soils and their clay content. (Based on data of the two sandy soils in the present study and Christensen, 1986.)

## Section 3. Activity of Microbes and the Microbial Biomass

### *Relation with Grazing Pressure of the Fauna*

Textural effects may impose physical restrictions on the ability of fauna to graze on microbes (Juma, 1993). Microcosm studies frequently have shown

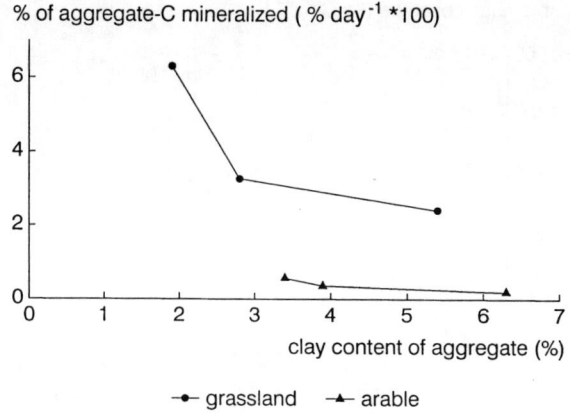

**Figure 6**  Relationship between the rate of C mineralization of organic C in aggregates of different classes (% day$^{-1}$ · 100) in sandy soils and their clay content. (Based on data of the two sandy soils in the present study.)

that the grazing activity of the fauna is higher in coarse-textured soils than in fine-textured soils and that grazing increases the activity of bacteria and enhances C mineralization (Clarholm, 1985; Coleman et al., 1978; Woods et al., 1982; Kuikman et al., 1990; Rutherford and Juma, 1992). Such stimulation suggests changes in rate-limiting factors such as availability of nutrients or energy (Anderson et al., 1981; Brussaard et al., 1991). Therefore we hypothesized that under field conditions the activity of microbes would be higher in coarse-textured soils than in fine-textured soils.

Hassink et al. (1993a) observed that bacteria constituted by far the largest biomass pool in Dutch grassland soils. They found a good correlation between the amount of bacterial biomass and the soil volume made up by pores with diameters between 0.2 and 1.2 μm, and between the biomass of nematodes and the soil volume made up by pores with diameters between 30 to 90 μm (Figures 7 and 8). The biomass of fungi and protozoa showed a poor correlation with any of the pore size classes.

It may be concluded that in sandy, as well as in loamy and clay soils, bacteria and nematodes are spatially separated. In sandy soils bacteria are probably protected from grazing by the fauna by their adsorption to, or coating by, clay and silt particles. It is likely that bacteria in loamy and clay soils are also adsorbed to or coated by clay and silt particles, but additionally are located within small pores within aggregates.

The ratio between the biomass of bacterivorous nematodes and the biomass of bacteria was greater in coarse-textured soils than in fine-textured soils, suggesting a higher grazing pressure of bacterivorous nematodes on bacteria in coarse-textured soils. For protozoa such a difference between soil types was

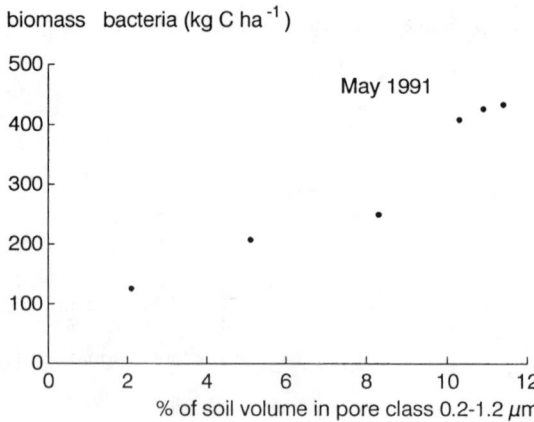

**Figure 7** Relationship between the biomass of bacteria and the percentage of soil volume enclosed by pores with diameters between 0.2 and 1.2 μm. (From Hassink et al., 1993a. *Soil Biol. Biochem.*, 25:47–55. © 1993 with kind permission of Elsevier Science Ltd., Kidlington OX51GB, UK.)

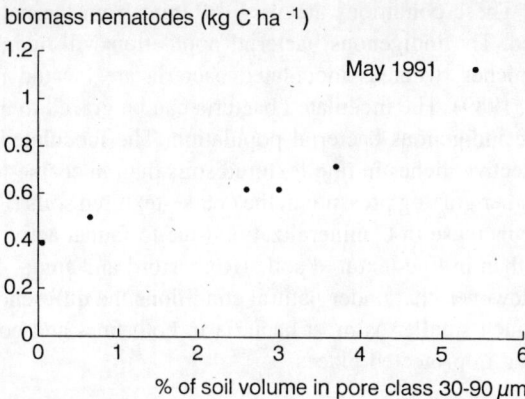

**Figure 8**  Relationship between the biomass of nematodes and the percentage of soil volume enclosed by pores with diameters between 30 and 90 μm. (From Hassink et al., 1993a. *Soil Biol. Biochem.*, 25:47–55. © 1993 with kind permission of Elsevier Science Ltd., Kidlington OX51GB, UK.)

not found. Contrary to the results of most microcosm studies, we found no correlation between grazing pressure (indicated as the ratio between biomass of bacterivorous nematodes and biomass of bacteria) and the activity of the bacteria (expressed as the amount of $CO_2$ produced per unit of bacterial biomass) (Figure 9) (Hassink et al., 1993a).

In one aspect our study is very different from the microcosm studies. In all microcosm studies bacteria and microfauna have been applied to previously

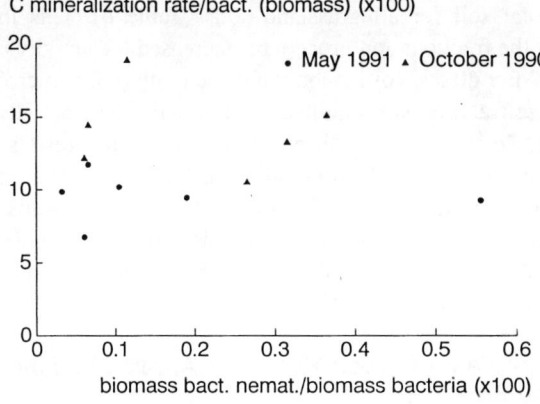

**Figure 9**  Relationship between the C mineralization rate (kg ha$^{-1}$ day$^{-1}$) per biomass of bacteria (kg ha$^{-1}$) and the biomass of bacterivorous nematodes (kg ha$^{-1}$) per biomass of bacteria (kg ha$^{-1}$). (From Hassink et al., 1993a. *Soil Biol. Biochem.*, 25:47–55. © 1993 with kind permission of Elsevier Science Ltd., Kidlington OX51GB, UK.)

sterilized soil. These conditions are very different from the natural situation that we studied. The indigenous bacterial population will mainly be present in protective niches, whereas inoculated bacteria are located in more open areas (Postma, 1989). The inoculated bacteria can be grazed to a much higher extent than the indigenous bacterial population. The inoculated bacteria can find more protective niches in fine-textured soils than in coarse-textured soils, leading to a higher grazing pressure in the coarse-textured soils (Postma, 1989) and a stronger increase in C mineralization (due to faunal activity) in coarse-textured soils than in fine-textured soils (Rutherford and Juma, 1992). It may be expected, however, that under natural conditions the differences in grazing pressure are much smaller as most bacteria in both fine- and coarse-textured soils are located in protected places.

### *Activity of Microbial Biomass in Different Primary Particles*

The activity of the microbial biomass was determined in the sand and clay + silt size fractions of the sandy, loamy, and clay grassland soils immediately after fractionation and after 60 days incubation of these fractions at 20°C; and in the arable sandy and clay soil in the clay + silt fraction immediately after fractionation. We expected that the activity of the microbial biomass would be lower in the clay + silt fraction than in the sand size fraction, as microbes and organic matter are more protected in the clay + silt fraction than in the sand size fraction and bacteria in the sand size fraction feed on more labile organic C (plant residues in various stages of decay) than bacteria in the clay + silt fraction. Immediately after fractionation, the activity of the microbial biomass in both the sand size fraction and the silt + clay fraction was higher than in the total soil for all grassland soils (Table 6). This indicates that disruption (by the fractionation procedure) increased the activity of the microbial biomass. After 60 days of incubation the activity of the microbial biomass in the sand size fraction was significantly lower than the activity in the clay + silt fraction. So in contrast with our expectations, our results show that a higher extent of physical protection did not decrease microbial activity. This activity was lower in the arable soils than in the grassland soils, while there was no consistent effect of soil texture. This indicates that the amount of available C is the main factor regulating the activity of the microbial biomass, and not soil texture and structure.

### Section 4. The Fate of Fresh Residues Applied to Fine- and Coarse-Textured Soils

It has frequently been observed that the accumulation of [14]C-labeled organic residues is directly related to the clay content of soils (Van Veen et al., 1985; Jocteur Monrozier et al., 1991; Amato and Ladd, 1992). Gregorich et al. (1991), however, found that clay content had no consistent effect on the amount of residual [14]C after application of glucose.

**Table 6   C Mineralization Rates (mg C kg$^{-1}$ day$^{-1}$ · 10$^2$) Divided by Amount of Microbial Biomass (mg C kg$^{-1}$) in Different Grassland and Arable Soils and in Their Silt + Clay (<20 μm) and Sand Size (>150 μm) Fractions During First 2 Weeks of Incubation and After 60 Days of Incubation at 20°C**

| | During first two weeks of incubation | | | After 60 days incubation | | |
| | | Fraction | | | Fraction | |
| Soil type | Total soil | <20 μm | >150 μm | Total soil | <20 μm | >150 μm |
|---|---|---|---|---|---|---|
| Grassland | | | | | | |
| Sand | 1.7 | 12.0 | 5.8 | 3.8 | 3.7 | 0.9 |
| Loam | 3.3 | 6.4 | 4.2 | 1.7 | 2.5 | 1.4 |
| Clay | 5.0 | 9.9 | 8.1 | 4.0 | 8.5 | 2.2 |
| Arable | | | | | | |
| Sand | 1.0 | 1.4 | n.d. | n.d. | n.d. | n.d. |
| Clay | 0.6 | 0.5 | n.d. | n.d. | n.d. | n.d. |

*Note:* n.d. = not determined.

To test whether differences in clay content and physical protection affect the amount of $^{14}C$ retained in the soil after a glucose application, we applied $^{14}C$-labeled glucose in solution to undisturbed field-moist; undisturbed dried and rewetted and dried, ground, and rewetted macroaggregates of a sandy soil and a clay, then determined the amount of $^{14}C$-$CO_2$ produced. All aggregates were incubated at –10 kPa at 20°C. We observed that part of the glucose penetrated the aggregates. It was hypothesized that the applied $^{14}C$ would be more protected inside the aggregates of the clay soil than of the sandy soil leading to a lower rate of $^{14}CO_2$ production. Grinding of the aggregates would decrease the extent of protection in the clay soil and increase $^{14}CO_2$ production while this would not be the case in the sandy soil.

We observed, however, that during the first 7 days the percentage of applied $^{14}C$ respired to $^{14}CO_2$ was higher in the undisturbed field-moist clay soil than in the field-moist sandy soil (Figure 10). Drying of the aggregates increased $^{14}CO_2$ evolution in the sandy soil, but not in the clay soil, and grinding did not increase $^{14}CO_2$ evolution in either of the soils. We observed no difference in $^{14}CO_2$ production between the sandy soil and clay after 17 and 33 days, in spite of the great differences in pore size distribution (much more open structure in the aggregates of the sandy soil; compare Figures 1 and 2), and the higher biomass of microfauna in the aggregates of the sandy soil.

This experiment and the results of Gregorich et al. (1991) show that the capacity of the soil to preserve applied C is not always directly correlated with soil texture or structure. There is a striking difference between the soils of Amato and Ladd (1992), Van Veen et al. (1985), and Jocteur Monrozier et al. (1991), who used the same soils, and the soils of Gregorich et al. (1991). For the soils of Gregorich a close correlation was observed between the clay content of the soil and its organic C content (Figure 11). This agrees with the statements that fine-textured soils have a higher capacity than coarse-textured soils to protect or preserve microbial biomass (Van Veen et al., 1985) and

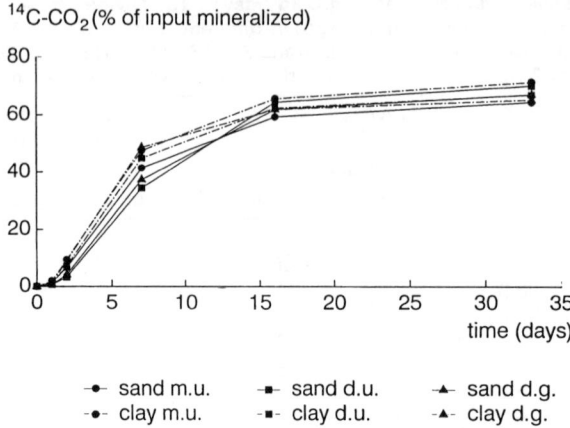

**Figure 10**    $^{14}C$ mineralization rate (expressed as percentage of $^{14}C$ mineralized) after applying $^{14}C$-labeled glucose to field-moist undisturbed (–10 kPa; m.u.), dried, and rewetted (to –10 kPa; d.u.) and dried, ground, and rewetted (to –10 kPa; d.g.) macroaggregates of a sandy soil and a clay soil.

organic C (Jenkinson, 1988). The coarse-textured soils of Amato and Ladd (1992) had similar organic C contents as the coarse-textured soils in the set of Gregorich. Their fine-textured soils, however, had considerably lower organic C contents than would have been expected assuming a linear relationship between clay content and protection capacity of a soil (Figure 11). For the relationship between clay content and microbial biomass the same kind of difference was found (Figure 12). This led us to the assumption that the fine-textured soils in the studies of Amato and Ladd (1992), Van Veen et al. (1985),

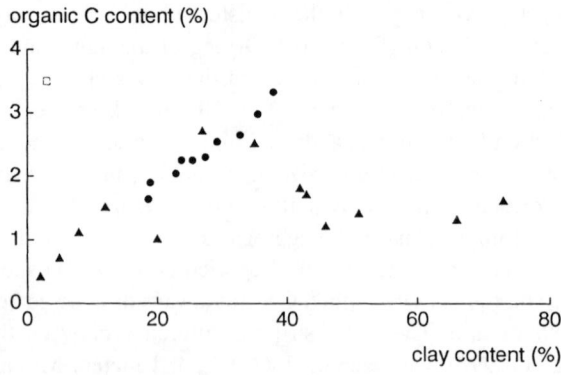

**Figure 11**    Relationship between the organic C and clay content (%) of arable soils. (Data from Gregorich et al., 1991; Amato and Ladd, 1992; and grassland soils in the present study.)

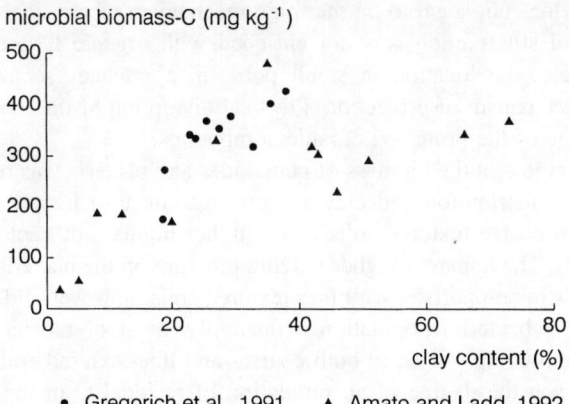

**Figure 12** Relationship between the amount of microbial biomass C (mg kg$^{-1}$) and the clay content of arable soils. (Data from Gregorich et al., 1991 and Amato and Ladd, 1992.)

and Jocteur Monrozier et al. (1991) had lower amounts of organic C and microbial biomass than they could preserve. Obviously, there were still sites available in their fine-textured soils where microbes and organic C could become physically protected. We assume that this was not the case for the soils from Gregorich et al. (1991). Considering this, we suggest that the accumulation of residual C is not always higher in fine-textured soils than in coarse-textured soils, but that this depends on the degree of saturation of the protective capacity of a soil. Fine- and coarse-textured soils in which the protective sites for organic C and microbes are occupied will retain less C than fine- and coarse-textured soils that have protective sites available.

The sandy grassland soil in our study had higher C contents than the arable soils of Gregorich et al. (1991). We observed, however, that the amount of C protected by clay particles was not different between arable and grassland soils (Figure 3). So we conclude that the extra amount of C present in the grassland soils was not physically protected. It has also been observed that the sandy grassland soils in The Netherlands contain charcoal (Hassink, 1994).

## CONCLUSIONS

The positive correlation between clay content and total organic C and microbial biomass indicates that the capacity to physically protect microbes and organics is greater in fine-textured soils than in coarse-textured soils. It may be concluded from our results that adsorption to or coating by clay and silt particles is the main mechanism of physical protection of organic matter and microbes in coarse-textured soils. The clay and silt fraction was highly enriched with organic C and microbial biomass. In fine-textured soils this mechanism does not seem to be the main mechanism of physical protection. The clay and silt fraction was not enriched with organic C and microbial biomass. Here, the location in small pores in aggregates seems to be an important mechanism of protection. Physical disruption of fine-textured soils released some of the protected organic compounds.

We observed that the biomass of nematodes and bacteria was regulated by the pore size distribution and clay content, leading to a higher biomass of nematodes in coarse-textured soils and a higher biomass of bacteria in fine-textured soils. The apparent higher grazing pressure on the bacteria in coarse-textured soils in comparison with fine-textured soils, however, did not lead to a more active bacterial population. Apparently, most of the bacteria were protected from being grazed in both coarse- and fine-textured soils.

In our view the degree of accumulation of residual C in the soil is not directly related to clay content, but depends on the degree of saturation of the protective capacity of a soil. Fine- and coarse-textured soils in which the protective sites for organic C and microbes are occupied will accumulate less C than fine- and coarse-textured soils that have protective sites available.

## REFERENCES

Amato, M. and Ladd, J. N., 1992. Decomposition of [14]C-labelled glucose and legume materials in soil: properties influencing the accumulation of organic residue C and microbial biomass C. *Soil Biol. Biochem.,* 24:455–464.

Anderson, R. V., Coleman, D. C., and Cole, C. V., 1981. Effect of saprophytic grazing on net mineralization, in *Terrestrial Nitrogen Cycles,* Clark, F. E. and Rosswall, T., Eds., *Ecological Bulletins,* 33:201–215.

Beare, M. H. and Bruce, R. R., 1993. A comparison of methods for measuring water-stable aggregates: implications for determining environmental effects on soil structure. *Geoderma,* 56:87–104.

Bottner, P., 1985. Response of microbial biomass to alternate moist and dry conditions in a soil incubated with [14]C- and [15]N-labelled plant material. *Soil Biol. Biochem.,* 17:329–337.

Brussaard, L., Kools, J. P., Bouwman, L. A., and de Ruiter, P. C., 1991. *Population Dynamics and Nitrogen Mineralization Rates in Soil as Influenced by Bacterial Grazing Nematodes and Mites,* Proceedings of the Xth International Soil Zoology Colloquium, Bangalore, India, August 7–13, 1988, pp. 517–523.

Chenu, C., 1993. Clay- or sand-polysaccharide associations as models for the interface between microorganisms and soil: water related properties and microstructure. *Geoderma,* 56:143–156.

Christensen, B. T., 1985. Carbon and nitrogen in particle size fractions isolated from Danish arable soils by ultrasonic dispersion and gravity-sedimentation. *Acta Agric. Scand.,* 35:175–185.

Christensen, B. T., 1986. Straw incorporation and soil organic matter in macro-aggregates and particle size separates. *J. Soil. Sci.,* 37:125–135.

Christensen, B. T., 1987. Decomposition of organic matter in particle size fractions from field soils with straw incorporations. *Soil Biol. Biochem.,* 19:429–435.

Christensen, B. T., 1992. Physical fractionation of soil and organic matter in primary particle size and density separates. *Adv. Soil Sci.,* 20:1–89.

Clarholm, M., 1985. Possible role for roots, bacteria, protozoa and fungi in supplying nitrogen to plants, in *Ecological Interactions in Soil: Plants, Microbes and Animals,* Fitter, A. H., Atkinson, D., Read, D. J., and Usher, M. B., Eds., Brit. Ecol. Soc. Spec. Publ., Oxford, pp. 355–365.

Coleman, D. C., Cole, C. V., Hunt, H. W., and Klein, D. A., 1978. Trophic interactions in soil as they affect energy and nutrient dynamics. I. Introduction. *Microb. Ecol.,* 4:345–349.

Dalal, R. C. and Mayer, R. J., 1986. Long-term trends in fertility of soils under continuous cultivation and cereal cropping in Southern Queensland. III. Distribution and kinetics of soil organic carbon in particle-size fractions. *Aust. J. Soil Res.,* 24:293–300.

Edwards, A. P. and Bremner, A. P., 1967. Microaggregates in soils. *J. Soil Sci.,* 18:64–73.

Elliott, E. T., 1986. Aggregate structure and carbon, nitrogen, and phosphorus in native and cultivated soils. *Soil Sci. Soc. Am. J.,* 50:627–633.

Elliott, E. T. and Coleman, D. C., 1988. Let the soil work for us, in *Ecological Implications of Contemporary Agriculture,* Eijsackers, H. and Quispel, A., Eds., Proc. 4th Eur. Ecol. Symp., Wageningen. *Ecological Bulletins,* 39:327–335.

Elliott, E. T., Anderson, R. V., Coleman, D. C., and Cole, C. V., 1980. Habitable pore space and microbial trophic interactions. *Oikos,* 35:327–335.

Elustondo, J., Angers, D. A., Laverdière, M. R., and N'Dayegamiye, A., 1990. Étude comparative de l'agrégation et de la matière organique associée aux fractions granulaométriques de spet sols sous culture de mais ou en prairie. *Can. J. Soil Sci.,* 70:395–402.

Foster, R. C., 1988. Microenvironments of soil microorganisms. *Biol. Fert. Soils,* 6:189–203.

Gregorich, E. G., Kachanoski, R. G., and Voroney, R. P., 1989. Carbon mineralization in soil size fractions after various amounts of aggregate disruption. *J. Soil Sci.,* 40:649–659.

Gregorich, E. G., Voroney, R. P., and Kachanoski, R. G., 1991. Turnover of carbon through the microbial biomass in soils with different textures. *Soil Biol. Biochem.,* 23:799–805.

Hassink, J., 1992. Effects of soil texture and structure on carbon and nitrogen mineralization in grassland soils. *Biol. Fert. Soils,* 14:126–134.

Hassink, J., Bouwman, L. A., Zwart, K. B., and Brussaard, L., 1993a. Relationships between habitable pore space, soil biota and mineralization rates in grassland soils. *Soil Biol. Biochem.,* 25:47–55.

Hassink, J., Bouwman, L. A., Zwart, K. B., Bloem, J., and Brussaard, L., 1993b. Relationships between soil texture, physical protection of organic matter, soil biota, and C and N mineralization in grassland soils. *Geoderma,* 57:105–128.

Hassink, J., 1994. Effects of soil texture and grassland management on soil organic C and N and rates of C and N mineralization. *Soil Biol. Biochem.,* 26:1221–1231.

Jenkinson, D. S., 1988. Soil organic matter and its dynamics, in *Russel's Soil Conditions and Plant Growth,* 11th ed., Wild, A., Ed., Longman, New York, pp. 564–607.

Jocteur Monrozier, L., Ladd, J. N., Fitzpatrick, R. W., Foster, R. C., and Raupach, M., 1991. Components and microbial biomass content of size fractions in soils of contrasting aggregation. *Geoderma,* 49:37–63.

Juma, N. G., 1993. Interrelationships between soil structure/texture, soil biota/soil organic matter and crop production. *Geoderma,* 57:3–30.

Kilbertus, G., 1980. Etude des microhabitats contenus dans les agrégats due sol. Leur relation avec la biomasse bactérienne et la taille des procaryotes présents. *Rev. Ecol. Biol. Sol,* 17:543–557.

Klute, A., 1986. Water retention: laboratory methods, in *Methods of Soil Analysis. I. Physical and Mineralogical Methods,* Klute, A., Ed., Agron Monogr 9, Am Soc Agron, Madison, Wisconsin, pp. 635–662.

Kortleven, J., 1963. Kwantitatieve aspecten van humusopbouw en humusafbraak. *Versl. Landb. Ond.,* nr. 69.1. PUDOC, Wageningen, 109 pp. (in Dutch).

Kuikman, P., Jansen, A. G., Van Veen, J. A., and Zehnder, A. J. B., 1990. Protozoan predation and the turnover of soil organic carbon and nitrogen in the presence of plants. *Biol. Fert. Soils,* 10:22–28.

Ladd, J. N., Amato, M., and Oades, J. M., 1985. Decomposition of plant material in Australian soils. III. Residual organic and microbial biomass C and N from isotope-labelled legume material and soil organic matter, decomposing under field conditions. *Aust. J. Soil Res.,* 23:603–611.

Ladd, J. N., Foster, R. C., and Skjemstad, J. O., 1993. Soil structure: carbon and nitrogen metabolism. *Geoderma,* 56:401–434.

Matus, F. M., 1997. Carbon and nitrogen distribution in aggregate size classes from a clay and sandy soil: relationship between habitable pore space, soil biota and mineralization rates. *Soil Biol. Biochem.,* in press.

Postma, J., 1989. Distribution and Population Dynamics of *Rhizobium* sp. Introduced into Soil, Thesis, Agric. Univ. Wageningen. 121 pp.

Postma, J. and Van Veen, J. A., 1990. Habitable pore space and population dynamics of *Rhizobium leguminosarum biovar trifolii* introduced into soil. *Microb. Ecol.,* 19:149–161.

Richter, J., Nuske, A., Habenicht, W., and Bauer, J., 1982. Optimized N-mineralization parameters of loess soils from incubation experiments. *Plant and Soil,* 68:379–388.

Rutherford, P. M. and Juma, N. G., 1992. Influence of texture on habitable pore space and bacterial-protozoan populations in soil. *Biol. Fert. Soils,* 12:221–227.

Spain, A. V., 1990. Influence of environmental conditions and some soil chemical properties on the carbon and nitrogen contents of some tropical Australian rain-forest soils. *Aust. J. Soil. Res.,* 28:825–839.

Tisdall, J. M. and Oades, J. M., 1982. Organic matter and water-stable aggregates in soils. *J. Soil Sci.,* 33:141–163.

Van Veen, J. A. and Paul, E. A., 1981. Organic carbon dynamics in grassland soils. I. Background information and computer simulation. *Can. J. Soil Sci.,* 61:185–201.

Van Veen, J. A. and Kuikman, P. J., 1990. Soil structural aspects of decomposition of organic matter. *Biogeochemistry,* 11:213–233.

Van Veen, J. A., Ladd, J. N., and Frissel, M. J., 1984. Modelling C and N turnover through the microbial biomass in soil. *Plant and Soil,* 76:257–274.

Van Veen, J. A., Ladd, J. N., and Amato, M., 1985. Turnover of carbon and nitrogen through the microbial biomass in a sandy loam and a clay soil incubated with [$^{14}$C(U)]glucose and [$^{15}$N] $(NH_4)_2SO_4$ under different moisture regimes. *Soil Biol. Biochem.,* 17:257–274.

Vance, E. D., Brookes, P. C., and Jenkinson, D. S., 1987. An extraction method for measuring soil microbial biomass. *Soil Biol. Biochem.,* 19:703–707.

Woods, L. E., Cole, C. V., Elliott, E. T., Anderson, R. V., and Coleman, D. C., 1982. Nitrogen transformation in soil as affected by bacterial-microfaunal interactions. *Soil Biol. Biochem.,* 14:93–98.

# CHAPTER 3

# Fungal and Bacterial Pathways of Organic Matter Decomposition and Nitrogen Mineralization in Arable Soils

M. H. Beare

## INTRODUCTION

The development of sustainable agricultural practices depends largely on promoting the long-term fertility and productivity of soils at economically viable levels. Efforts to achieve these goals have focused on (1) lowering fertilizer inputs in exchange for a higher dependence on biologically fixed and recycled nutrients, (2) reducing pesticide uses while relying more on crop rotations and biocontrol agents, (3) decreasing the frequency and intensity of tillage, and (4) increasing the return of crop residues and animal wastes to land. The principal objectives of these approaches are to match the supply of soil nutrients with the fertility demands of the crops, to maintain acceptable pest tolerance levels, and to develop soil physical properties that optimize oxygen supply, water infiltration, and water-holding capacity at levels that minimize the losses of nutrients by leaching and gaseous export. Determining the suitability of these "sustainable" practices to a broad range of crops, soil types, and climatic regimes requires an understanding of their effects on the physical, chemical, and biological properties of soils.

The importance of soil biota as causal mechanisms for sustaining the fertility and productivity of soils has been the focus of several major programs on the ecology of arable farming systems. These include, but are not restricted to, (1) the long-term conventional (CT) and no-tillage (NT) trials (Horseshoe Bend site) of the Georgia Agroecosystems Project in the United States (Stinner et al., 1984; Hendrix et al., 1986; Beare et al., 1992), (2) the conventional and integrated farming trials (Lovinkhoeve site) of the Dutch Programme on soil Ecology of Arable Farming Systems (Brussaard et al., 1988; Kooistra et al., 1989), (3) the barley, grass ley, and lucerne ley trials (Kjettslinge site) of the Swedish project on the Ecology of Arable Lands (Andrén et al., 1990), (4) the long-term stubble mulch and no-tillage trials (Akron site) at the Central Great Plains Research Station in the United States (Elliott et al., 1984; Holland and Coleman, 1987), and (5) the cultivated barley trials (Ellerslie and Breton sites) of the University of Alberta, Canada (Rutherford and Juma, 1989).

Within these programs much attention has been directed at understanding the contributions of fungal- and bacterial-based food webs to the accumulation and loss of soil organic matter (SOM) and to nutrient cycling. The importance of distinguishing these two primary pathways is based on the theses that (1) bacteria have lower C assimilation efficiencies and faster turnover rates than fungi, factors that are likely to increase rates of nutrient mineralization and organic matter decomposition in bacterially dominated soils, and that (2) the mycelial growth form is more conservative of energy and nutrients, enhancing organic matter storage and nutrient retention in fungal dominated soils (Adu and Oades, 1978; Paustian, 1985; Holland and Coleman, 1987).

Where bacterial production is greater, bacterial-feeding fauna are expected to dominate. The most common bacterial-feeding fauna are protozoa (Clarholm, 1981; Laybourn-Parry, 1984) and many nonstylet-bearing nematodes (Sohlenius et al., 1987), which require water films for locomotion and feeding. They are generally believed to increase organic matter loss and nutrient mineralization due to their relatively large biomass and high turnover rates (Stout, 1980; Kuikman and van Veen, 1989) and to their feeding on bacteria (Coleman et al., 1984). In fungal-dominated soils, fungal-feeding fauna such as many non-plant-parasitic stylet-bearing nematodes (Parmelee and Alston, 1986) and various microarthropod groups (Walter, 1987; Mueller et al., 1990) are expected to be more important. In arable soils, fungal-feeding fauna usually comprise a smaller biomass and have slower turnover rates than bacterial-feeding fauna; factors that are expected to reduce their direct contributions to organic matter decomposition. However, fungal-feeding microarthropods can also enhance residue decomposition rates through their stimulations of fungal growth (Santos and Whitford, 1981) or by direct comminution of substrates (Seastedt, 1984). Low to moderate levels of grazing can stimulate fungal production and, thus, fungal immobilization of nutrients, whereas, high levels of grazing tend to increase nutrient mineralization (Hanlon and Anderson, 1979; Beare et al., 1992).

In many of the aforementioned studies differences in the structure and function of soil food webs were proposed to explain their differences in organic matter dynamics and nutrient cycling. Several authors, for example, have proposed that cultivation of soils by ploughing favors organisms with short generation times, small body size, rapid dispersal, and generalist feeding habits (Andrén and Lagerlöf, 1983; Ryszkowski, 1985). Based on these observations, Hendrix et al. (1986) hypothesized that the predominance of fungal- and bacterial-based food webs in NT and CT agroecosystems, respectively, could account for many of their observed differences in organic matter turnover and nutrient cycling. In several studies of arable soils data on the abundance and biomass of microflora and fauna have been used to estimate the flows of C and N through the soil food webs (e.g., Hendrix et al., 1987; Brussaard et al., 1990; Moore et al., 1990; Paustian et al., 1990; Beare et al., 1992; Didden et al., 1994). Other have used experimental manipulations in the field and laboratory to investigate how the trophic interactions in fungal- and bacterial-based food webs influence rates of organic matter turnover and nutrient mineralization (e.g., Parmelee et al., 1990; Mueller et al., 1990; Beare et al., 1992). This chapter elaborates on the above-mentioned reports, adding new information and giving special attention to the importance of fungal and bacterial pathways in regulating residue decomposition, nutrient mineralization, and the storage of SOM. Though many of the examples cited here come from studies of the long-term CT and NT plots at the Horseshoe Bend (HSB) site, I have attempted to compare and contrast these findings with those of other sites, wherever possible. The principal objectives of this review are (1) to identify some of the primary ways that soil cultivation affects the structure and function of soils food webs and (2) to distinguish some of the mechanisms by which fungal- and bacterial-based food webs regulate soil processes so that they might be better managed to sustain the fertility and productivity of arable lands.

## BELOWGROUND FOOD WEBS

Estimating the contributions of fungi and bacteria to the transformations of energy and matter in soils is made more difficult by the complexity of their interactions with other organisms in the belowground food web (Coleman, 1985) and by the spatial and temporal heterogeneity of their activities (Anderson, 1988). One of the more common approaches to evaluating the relative contributions of soil biota to heterotrophic processes involves budgeting their biomass, production, and respiration in accordance with their functional classification in soils. Biomass C and N budgets for belowground food webs of arable soils have been described in various reports (e.g., Hendrix et al., 1987; Brussaard et al., 1990; Paustian et al., 1990; Andrén et al., 1990; Zwart et al., 1994). Results from four of these studies are summarized in Table 1 (after Brussaard et al., 1990) including more recent and comprehensive data from the HSB and Lovinkhoeve sites (see Tables 1 and 2 for sources of data).

The original C-budget estimates cited by Brussaard et al. (1990) for the HSB site (Hendrix et al., 1987) were based on a relatively small dataset collected under a cool season (winter/spring) winter rye crop. The original findings grossly underestimated the biomass of fungi due, in part, to the incomplete extraction of fungal hyphae and computational errors in estimating their population densities. The data presented here (Table 1) for fungi, bacteria, protozoa, and nematodes were calculated from results presented by Hendrix

Table 1 Biomass (kg C ha⁻¹) of Microbial and Faunal Groups as Percentage of Total Organism Biomass in Agricultural Soils from Four Different Arable Land Projects

| | Horseshoe Bend[a] | | Lovinkhoeve[b] | | Kjettslinge[c] | | | | Ellerslie[d] | Breton[d] |
| | NT | CT | CF | IF | B0 | B120 | GL | LL | CT | CT |
| | | | | | % of total biomass | | | | | |
|---|---|---|---|---|---|---|---|---|---|---|
| Bacteria | 48.1 | 56.3 | 94.0 | 75.0 | 30.0 | 27.7 | 34.6 | 32.1 | **75.3** | **47.6** |
| Fungi | 46.3 | 39.9 | 0.86 | 0.97 | 64.4 | 70.7 | 61.5 | 64.3 | 24.6 | 52.3 |
| Protozoa | 1.44 | 2.18 | 4.90 | 5.90 | 4.72 | 1.05 | 2.92 | 1.57 | 0.01 | 0.00 |
| Nematodes | 0.073 | 0.114 | 0.24 | 0.26 | 0.07 | 0.03 | 0.07 | 0.08 | 0.06 | 0.03 |
| Microarthropods | 0.080 | 0.027 | 0.24 | 0.15 | 0.02 | 0.02 | 0.02 | 0.03 | n.d. | n.d. |
| Macroarthropods | 0.010 | 0.030 | n.d. | n.d. | 0.04 | 0.03 | 0.08 | 0.23 | n.d. | n.d. |
| Enchytraeids | 0.339 | 0.268 | 0.18 | 0.07 | 0.17 | 0.10 | 0.04 | 0.10 | n.d. | n.d. |
| Earthworms | 3.68 | 1.17 | 0.00 | 17.6 | 0.47 | 0.42 | 0.78 | 1.58 | | |
| Total biomass C (kg C ha⁻¹) | 1,630 | 1,793 | 241 | 326 | 2,338 | 3,254 | 2,602 | 2,801 | 609 | 554 |

Note: n.d. = not determined; CT = conventional tillage; NT = no-tillage; CF = conventional farming; IF = integrated farming; B0 = Barley, 0 kg N fertilizer/ha; B120 = Barley, 120 kg N fertilizer/ha; LL = lucerne ley; GL = fescue grass ley; bold numbers are totals.

a Horseshoe Bend Experimental Area, GA, United States. Hiwassee sandy clay loam, Rhodic Kanhapludult, 0 to 21 cm (except where noted otherwise), annual average (monthly sampling). Sources of data as described in Table 3.

b Lovinkhoeve site, The Netherlands. Typic Fluvaquent, silt loam, 0 to 25 cm, winter wheat, spring/summer samples (Zwart et al., 1994).

c Kjettslinge site, Sweden. Mixed, frigid Haplaquoll, loam, 0 to 27 cm, barley, September 1982–1983 (Paustian et al., 1990; Andrén et al., 1990).

d Ellerslie site, Alberta, Canada, Black Chernozem, silt clay loam, 0 to 10 cm, barley, summer/autumn sampling; and Breton site, Alberta, Canada, Gray Luvisol, silt loam, 0 to 10 cm, barley, summer/autumn sampling (Rutherford and Juma, 1989).

Updated from Brussaard, L. et al., 1990. Neth. J. Agric. Sci., 38:283–302.

**Table 2** Seasonal Differences in Biomass (kg C ha⁻¹) of Microbial and Faunal Groups in Conventional Tillage and No-Tillage Soils at HSB

| | Summer–Autumn | | | Winter–Spring | | | |
|---|---|---|---|---|---|---|---|
| | NT | CT | NT:CT ratio | NT | CT | NT:CT ratio | Season p <0.05 |
| Fungi[a] | | | | | | | |
| Biomass | 799 | 740 | 1.08 | 711 | 690 | 1.03 | ** |
| % Total | 49.5 | 39.9 | | 43.2 | 39.9 | | |
| Bacteria[a] | | | | | | | |
| Biomass | 751 | 1053* | 0.71 | 816 | 968* | 0.84 | |
| % Total | 46.5 | 56.7 | | 49.6 | 56.0 | | |
| F:B ratio | 1.06 | 0.70* | | 0.87 | 0.71* | | |
| Protozoa[a] | | | | | | | |
| Biomass | 33 | 52* | 0.63 | 14 | 26* | 0.54 | ** |
| % Total | 2.04 | 2.80 | | 0.85 | 1.50 | | |
| Nematodes | | | | | | | |
| Fungivore[a] | | | | | | | |
| Biomass | 0.18 | 0.75* | 0.24 | 0.09 | 0.17 | 0.53 | |
| % Total | 0.011 | 0.040 | | 0.005 | 0.010 | | |
| Bacterivore | | | | | | | |
| Biomass | 0.87 | 1.25 | 0.70 | 0.77 | 1.29 | 0.60 | |
| % Total | 0.054 | 0.067 | | 0.047 | 0.075 | | |
| Omn-Pred. | | | | | | | |
| Biomass | 0.06 | 0.07 | 0.86 | 0.07 | 0.06 | 1.17 | |
| % Total | 0.004 | 0.004 | | 0.004 | 0.003 | | |
| Microarthropods[b] | | | | | | | |
| Biomass | 1.72 | 0.68* | 2.53 | 0.90 | 0.30 | 3.00 | |
| % Total | 0.107 | 0.037 | | 0.055 | 0.017 | | |
| Enchytraeids[c] | | | | | | | |
| Biomass | 3.60 | 2.91* | 1.24 | 7.50 | 6.66* | 1.13 | ** |
| % Total | 0.220 | 0.159 | | 0.456 | 0.385 | | |
| Earthworms[d] | | | | | | | |
| Biomass | 25.0 | 5.8* | 4.31 | 95.0 | 36.0* | 2.64 | ** |
| % Total | 1.55 | 0.31 | | 5.77 | 2.08 | | |

[a] Calculated from data of Hendrix et al. (1989), and Beare et al. (unpublished), monthly sampling, 0 to 21 cm; summer–autumn — July 1986 to Nov. 1986; winter–spring — Dec. 1986 to July 1987. Asterisks indicate significant effects of tillage within season (*, $p < 0.05$, $t$-test) and season across tillages (**, ANOVA, $p < 0.05$).

[b] Summer–autumn values were calculated from data of House and Parmelee (1985), ≈ monthly sampling, 0 to 5 cm, May 1983 to Dec. 1983. Values for winter–spring were estimated to be ≈ 50% of the summer–autumn values, based on data of House and Parmelee (1985), Parmelee et al. (1990), and Beare et al. (1992).

[c] Calculated from data of von Vliet et al. (1994), monthly sampling, 0 to 15 cm, Jan. 1991 to Jan. 1993.

[d] Calculated from data of Parmelee et al. (1990), ≈ monthly sampling, 0 to 15 cm, Jan. 1986 to April 1987.

et al. (1989) and those of Beare (unpublished). Measures of enchytraeid biomass were also underestimated (Parmelee et al., 1990) and are replaced here with more recent data collected using a higher efficiency extraction technique (van Vliet et al., 1995). The earthworm data were taken from a more comprehensive analysis of their population dynamics (Parmelee et al., 1990). Other than for micro- and macro-arthropods (House and Parmelee, 1985), the updated data also represent annual averages of regular samplings (approximately monthly) taken throughout the year rather than those of a single season.

These updated findings present an interesting contrast to those of the other sites (Table 1). At HSB, bacterial biomass was approximately 1.4 times greater than fungal biomass in CT (0 to 21 cm), whereas fungal and bacterial biomass were nearly equal in NT. This difference between tillages contrasts markedly with the clear dominance of bacterial biomass under both conventional (CF) and integrated (IF) farming practices at the Lovinkhoeve site and the much greater fungal biomass in the barley (especially fertilized barley [B120]) and ley treatments at the Kjettslinge site. Excluding the Canadian sites, protozoa made up the highest percentage of total biomass at the Lovinkhoeve site where the microbial biomass was composed almost entirely of bacteria. The highest biomass of protozoa, however, was recovered from the Kjettslinge site where the biomass of bacteria was much lower than that of fungi. Still, of the four treatments at this site, unfertilized barley (B0) soils had the highest biomass of bacterial feeding protozoa and the lowest biomass of bacteria. Notably, the relative biomass of protozoa at the two Canadian sites (Ellerslie and Breton) was many times higher than those of the other sites, comprising 25 to 52% of the total heterotrophic biomass. This may be due to the fact that the samples were collected during a very wet summer and that both cystic and active forms of protozoa were included in the population estimates. The somewhat higher biomass of protozoa and nematodes (60% bacterivores) (Table 2) in CT soils at HSB was consistent with the higher biomass of bacteria as compared with NT. Microarthropods (dominated by fungivorous and omnivorous Collembola) made up the highest percentage of total biomass at the Lovinkhoeve site where 98 to 99% of the microbial biomass was composed of bacteria. At Kjettslinge, microbial biomass generally decreased (B0 < GL < LL < B120) as the biomass of soil fauna increased (B120 < LL $\cong$ GL < B0) across the treatments. A similar difference was found at HSB, where the higher biomass of fauna in NT (91 kg C ha$^{-1}$) as compared to CT (68 kg C ha$^{-1}$) corresponded with a significantly lower microbial biomass. In contrast, although soil fauna (especially protozoa and earthworms) made up a much higher percentage of the total biomass in IF (24%) as compared to CF (5.1%) soils, this difference was not reflected in the microbial biomass of IF (248 kg C ha$^{-1}$) and CF (230 kg C ha$^{-1}$) soils.

These apparent inconsistencies in the size and composition of decomposer food webs emphasize the need to better understand the impact of management on biomass specific rates of activity (e.g., consumption, respiration, mineralization) in each of the principal functional groups. Furthermore, the biomass of organisms varies both spatially and temporally, as do the processes of nutrient mineralization and immobilization and organic-matter decomposition, which they mediate. Understanding these variances is essential to designing sustainable cropping practices that match the supply of nutrients and water with the demands of the plant.

## Spatial Variation

Spatial variation in soils occurs both vertically and horizontally. The vertical stratification of physical, chemical, and biological properties is inherent to most native soils. In arable soils, however, vertical stratification is significantly altered by management factors such as tillage, fertilizer placement, and irrigation. These alterations may have significant consequences for the storage and loss of organic matter and nutrients.

Taken as an annual average, the vertical distribution of fungi and bacteria at HSB was strongly influenced by tillage (Figure 1A). In NT, fungal biomass was concentrated near the soil surface (0 to 5 cm), decreasing much more significantly with depth than the bacterial biomass. Fungal and bacterial biomass remained relatively constant in the plough layer (0 to 13 cm) of CT, but were much lower at the greatest depth (13 to 21 cm). In NT, the microbial biomass shifted from one dominated by fungi near the soil surface (F:B ratio = 1.40) to one dominated by bacteria below 13 cm (F:B ratio = 0.57) (Figure 1B). Though the biomasses of fungi and bacteria were similar near the soil surface in CT, bacteria dominated the microbial biomass at greater depths. Vertical changes in composition of the microbial community (as measured by F:B ratios) in CT were much less pronounced than in NT. These patterns of vertical stratification are somewhat different from those reported by Doran (1980) in which most groups of aerobic and anaerobic microorganisms were more abundant in the surface soil (0 to 7.5 cm) of NT than CT, with the reverse being true deeper in the plough layer (7.5 to 15 cm).

Viewing the stratification of organisms with respect to the distributions of organic matter in soils can help to shed light on the mechanisms of organic matter turnover and to explain site- or management-specific differences in its accumulation or loss. For example, as a percentage of whole soil C, bacterial biomass increased with increases in sample depth, irrespective of tillage at HSB (Figure 1C). Despite the similar pattern, bacterial biomass in surface soils (0 to 13 cm) of CT comprised a much higher percentage of whole soil C (2.8 to 3.9%) than the bacterial biomass in NT (1.7 to 2.8%). In NT, fungal biomass remained ≈2.3% of total soil C (kg ha$^{-1}$ cm$^{-1}$) at each depth. Similarly, fungal biomass in CT was between 2.5 and 2.9% of total C in the surface soils, being slightly lower at the deepest depth. Overall, the total microbial biomass in NT comprised ≈4.8% of the whole soil C (0 to 21 cm). In CT, however, the total microbial biomass made up ≈6.1% of the whole soil C, a higher proportion of which was bacterial in origin and concentrated in the plough layer (≈ 0 to 13 cm) where it is susceptible to the perturbations imposed by tillage and the more extreme dry/wet cycles of bare soil surfaces.

Doran (1980) argued that the less oxidative condition of NT soils would reduce rates of N mineralization and nitrification while enhancing denitrification as compared with CT soils. Though supported by some studies (Rice and

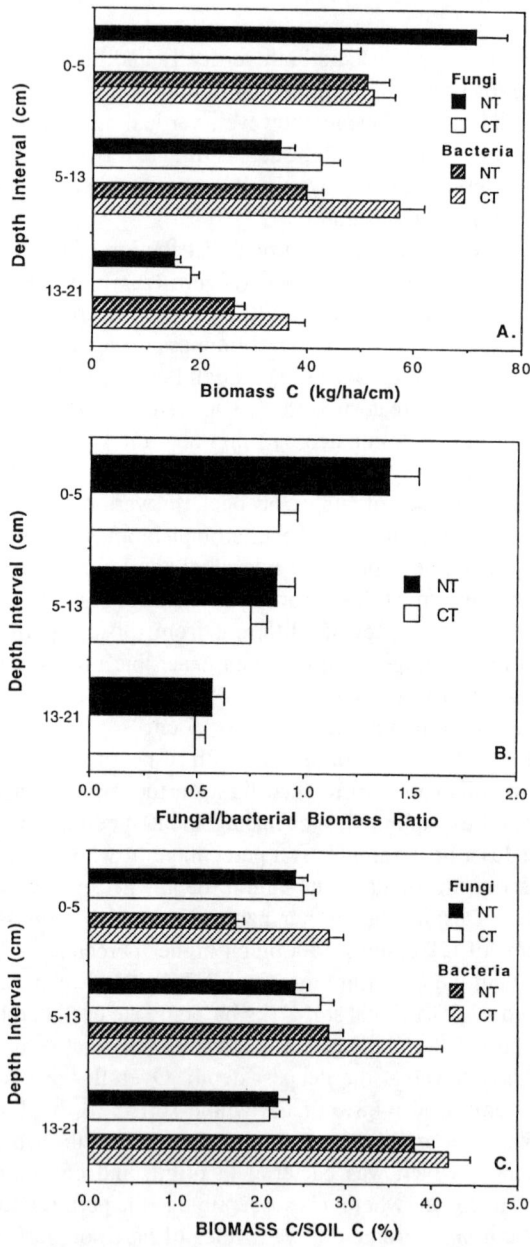

**Figure 1** The vertical stratification of (**A**) fungal and bacterial biomass (kg C ha$^{-1}$ cm$^{-1}$), (**B**) fungal-to-bacterial biomass ratios, and (**C**) their biomass as a percentage of whole soil C at each depth in CT and NT soils at HSB. (From Beare, M. H., unpublished.)

Table 3  Seasonal Differences in the Vertical Stratification of Mineral N (kg ha$^{-1}$ cm$^{-1}$) in Conventional Tillage (CT) and No-Tillage (NT) Soils at the HSB

| | Summer–Autumn | | | Winter–Spring | | | |
| Depth | NT | CT | Ratio NT:CT | NT | CT | Ratio NT:CT | Season $p < 0.05$ |
|---|---|---|---|---|---|---|---|
| 0–5 | 2.28 | 2.50 | 0.91 | 0.74 | 0.80 | 0.93 | ** |
| 5–13 | 1.31 | 2.01* | 0.66 | 0.48 | 0.53 | 0.90 | ** |
| 13–21 | 1.19 | 1.68* | 0.71 | 0.59 | 0.60 | 0.98 | ** |
| 0–21 | 1.50 | 2.00* | 0.75 | 0.58 | 0.65 | 0.94 | ** |
| NO$_3$-N (% total) | 84 | 89 | | 65 | 67 | | |

Note: Calculated from data of Hendrix et al. (1989) and Beare et al. (unpublished), monthly sampling, 0 to 21 cm; summer–autumn — July 1986 to Nov. 1986 (n = 6); winter–spring — Dec. 1986 to June 1987 (n = 6). Stars indicate significant effects of tillage within season (*, $p < 0.05$, t-test) and season across tillages (**, ANOVA, $p < 0.05$).

Smith, 1983; Aulakh et al., 1984), not all results from HSB are in total agreement with this hypothesis. The average mineral N content of CT soils was significantly higher than that of NT soils (Table 3), though there were strong seasonal differences (discussed below). The greater vertical stratification of mineral N in NT as compared with CT soils is consistent with their differences in the distribution of microorganisms (Figure 1) and fauna (Hendrix et al., unpublished). Furthermore, the much higher concentrations of mineral N at depth in CT may be responsible for the higher NO$_3$ leaching losses found in these soils (Stinner et al., 1984). Other studies at HSB indicate that nitrification and denitrification activities are both higher in the surface soil (0 to 5 cm) of NT as compared with CT, with the reverse pattern being observed at greater depths (5 to 21 cm) (Groffman, 1985). However, when totaled over the top 21 cm, there were no differences in their nitrification and denitrification activities on an annual basis.

Brussaard et al. (1990) also reported differences in the distribution of organisms in the CF and IF soils of The Netherlands. In IF soils, where tillage was shallow without inversion, the biomass of microbes and bacterivorous and fungivorous nematodes was higher in the top 10 cm, whereas the reverse was generally true for CF soils where crop residues were inverted by deeper tillage. Rates of *in situ* N mineralization and O$_2$ consumption were also higher in IF than CF soils and concentrated near the soil surface, consistent with distributions of bacteria and bacterivorous fauna (Bloem et al., 1994). Similarly, a marked stratification of microbial and faunal populations was also noted in the lucerne and grass ley trials at the Kjettslinge site, though little or no stratification was found associated with the cultivated barley soils (Andrén et al., 1990; Sohlenius et al., 1987).

Though much better described for soil physical and chemical properties (e.g., Jackson and Caldwell, 1993), soil organisms and rates of biologically meditated processes also tend to have highly skewed horizontal distributions in soils. For example, using a geostatistical approach, Robertson and Freckman (unpublished, as cited in Robertson, 1994) have shown that 40 to 60% of the variance in populations of bacterivorous, fungivorous, and omnivorous nematodes from a soybean field in Michigan were spatially dependent at scales of 2 to 75 m. Microbial populations and rates of biologically mediated processes (e.g., N mineralization, denitrification, etc.) are also found to be spatially dependent at these scales (Robertson et al., 1988; Parkin, 1987).

As such, understanding the vertical and horizontal distribution of fungal- and bacterial-mediated processes, the nature and extent of their trophic coupling at similar spatial scales, and the soil properties that determine them may contribute significantly to adapting spatially sensitive, variable-input farm equipment to soil-specific farming strategies (Robertson, 1994).

## Temporal Variation

Understanding how soil biota respond to seasonal and, hence, climatological variation in the soil environment can also help in designing sustainable crop production practices. In many cases problems of nutrient supply, pest control, or water management can be attributed to critical periods (e.g., seasons) within the cropping cycle. Determining how factors such as the timing and placement of crop residues, fertilizers, and irrigation water influence soil biotic activity will be important to adopting practices that minimize these constraints to sustainable crop production.

To illustrate this point, the HSB data discussed above were summarized by cropping seasons, where some very notable differences emerged (Table 2). Whereas the total biomass of fungi and bacteria in each of NT and CT did not differ tremendously between cropping seasons, ratios of fungal-to-bacterial biomass (F:B) revealed some tillage-dependent seasonal shifts in the composition of the microbial community. Soil F:B ratios were very similar under the summer/autumn (warm season) and winter/spring (cool season) crops in CT; however, the F:B ratio was much higher in the warm-season than the cool-season soil of NT. Fungal and bacterial biomass tended to be higher in the warm-season than in the cool-season soils of CT. In NT, however, fungal biomass was higher and bacterial biomass lower in the warm-season soils. Using the conversion factors presented in Table 4, bacteria were estimated to account for 63 and 73% of the total annual heterotrophic respiration in NT and CT, respectively (Table 5). Furthermore, while fungal and bacterial respiration in CT remained a relatively constant percentage of the total in both seasons, microbial contributions to respiration in NT were nearly 10% higher in the warm season than the cool season.

Broadly speaking, the NT:CT biomass ratios for each of the soil faunal groups reflect their anticipated functional relationship with the primary decomposer groups (bacteria and fungi). For example, the biomass of bacterial-feeding fauna (protozoa and bacterivorous nematodes) was considerably lower in NT than CT in both seasons, a result that is consistent with the lower

**Table 4    Conversion Factors Used in Calculating the Average Annual Biomass C (B) and Annual Production (P) and Respiration (R) of Each of the Major Organismal Groups in Soils at HSB**

| Organismal group | Biomass C conversion factors | P:B ratio[a] | R:P ratio[a] |
|---|---|---|---|
| Bacteria[b] | 118 µg C/$10^9$ cells (soil) | 4.0 | 1.5 |
| | 176 µg C/$10^9$ cells (litter) | | |
| Fungi[c] | 0.882 µg C/m hyphae (soil) | 1.5 | 1.5 |
| | 0.141 µg C/m hyphae (litter) | | |
| Protozoa[d] | | | |
| Ciliates | 6.3 µg C/$10^4$ cells | | |
| Amoebae | 4.5 µg C/$10^4$ cells | 4.0 | 2.0 |
| Flagellates | 1.2 µg C/$10^4$ cells | | |
| Nematodes[e] | | | |
| Bacterivore | 0.064 µg C/individual | | |
| Fungivore | 0.031 µg C/individual | 6.0 | 1.67 |
| Omni-Predator | 0.049 µg C/individual | | |
| Microarthropods[f] | 5.0 µg C/individual | 3.0 | 2.0 |
| Macroarthropods | 50% of AFDW | n.d. | n.d. |
| Enchytraeids[g] | 50% of AFDW | 2.0 | 3.5 |
| Earthworms[h] | 50% of AFDW | 4.0 | 3.5 |

[a] Sources of these values were Clarholm (1985), Persson et al. (1980), and Hendrix et al. (1987).
[b] Avg. biovolume = 0.8 µm³/cell; corrections for density, dry weight, and C content after Bakken and Olson (1983).
[c] Avg. hyphal diameter = 2.75 µm; density, corrections for dry weight, and C content after van Veen and Paul (1979).
[d] Calculated assuming dry masses of 1.4, 1.0, and 0.26 ng per ciliate, amoeba, and flagellate, respectively (see Beare et al., 1992, for primary references).
[e] After Freckman and Mankau (1986) and Golebiowska and Ryszkowski (1977).
[f] After Peterson and Luxton (1982).
[g] Primary ash-free dry weight (AFDW) data from van Vliet et al. (1994).
[h] Primary AFDW data from Parmelee et al. (1990).

biomass of bacteria in NT soils. Furthermore, in NT soils, where fungi make a relatively larger contribution to the total microbial biomass, the biomass of microarthropods (dominated by fungal-feeding Oribatida, Prostigmata and uropodid Mesostigmata mites and Collembola) was two- to threefold higher than in CT.

Protozoa are generally considered to be the most important consumers of bacteria in soils (Clarholm, 1981). Irrespective of tillage, naked amoebae (65%), followed by flagellates (32%) and ciliates (2.5%), composed the highest percentage of the protozoan biomass at HSB. The biomass of protozoa was highest in the warm season, constituting ≈33 and 52 kg C ha$^{-1}$ in NT and CT, respectively; nearly double their biomass in the cool season. In spite of this, no seasonal differences in bacterial biomass were noted. Assuming steady-state conditions, a turnover of 4 yr$^{-1}$ and a C yield of 50%; protozoa were estimated to consume ≈282 and 468 kg C ha$^{-1}$ yr$^{-1}$ in NT and CT, respectively. Separated by season, the consumption of C by protozoa amounted to ≈13 and 15% of bacterial production in the warm season and 5 and 8% of bacterial production in the cool season, for NT and CT, respectively.

That protozoa can stimulate the mineralization and plant uptake of N and P (Elliott et al., 1979; Clarholm, 1985; and Kuikman and van Veen, 1989) is well known. Assuming a C:N ratio of 4 for bacteria and 7 for protozoa (Brussaard et al., 1990), protozoa at HSB were estimated to mineralize ≈54 and 90 kg N ha$^{-1}$ yr$^{-1}$ in NT and CT, respectively. More than 65% of their contribution to N mineralization could be attributed to the warm season in both tillages. In contrast, protozoa were estimated to mineralize ≈30 and 43 kg N ha$^{-1}$ yr$^{-1}$ in CF and IF soils, respectively at the Lovinkhoeve site (Didden et al., 1994). Their contribution to N-flux in the fertilized barley treatments at the Kjettslinge site in Sweden was ≈30 kg N ha$^{-1}$ yr$^{-1}$, which amounted to 16% of the total N-flux in this treatment (Andrén et al., 1990). As a matter of comparison, the biomass of bacterivorous nematodes at HSB was less than 5% of the protozoan biomass. They also contributed <0.1% of the total heterotrophic respiration and consumed <0.5% of the bacterial production in both NT and CT.

Nematodes and microarthropods are generally believed to be the dominant fungal-feeders in most arable soils. At HSB, the microarthropod community is dominated by fungivorous Oribatida and Prostigmata mites and collembola (Mueller et al., 1990); their biomass in NT averaging 2.5 to 3.0 times that of CT soils. In contrast, the largest biomass of fungal-feeding nematodes was found in the warm-season soils of CT. Relatively high populations of nematodes are often found in association with incorporated residues in CT where the biomass of fungi can be 2 to 3 times higher than that of surface residues in NT (Beare et al., 1992). However, in difference to their anticipated trophic links, the biomass of fungal-feeding nematodes was much higher in CT than NT soils relative to the difference in fungal and bacterial biomass. Assuming an assimilation efficiency of 0.6 and the values in Table 4, fungivorous nematodes were estimated to consume 2.7 and 9.2 kg C ha$^{-1}$ yr$^{-1}$ in NT and CT, respectively, each amounting to <1.0% of the fungal production, regardless of tillage. Assuming C:N ratios of 10 and 10 for fungi and nematodes, respectively, the contributions of fungivorous nematodes to N mineralization were estimated to be <0.2 kg N ha$^{-1}$ yr$^{-1}$. Similarly, due to their generally low biomass, slow turnover rates and relatively low assimilation efficiencies, microarthropods were estimated to contribute very little to the C and N flux in the bulk soils of CT and NT at HSB, a finding that is consistent with observations at the Lovinkhoeve (Brussaard et al., 1990; Didden et al., 1994) and Kjettslinge sites (Andrén et al., 1990).

Although the preliminary findings of Hendrix et al. (1987) indicated that earthworms comprised ≈14% of the total biomass C in NT soils, the somewhat more comprehensive data presented here indicate that 3 to 4% may represent a more reasonable estimate of their contribution on an annual basis. Earthworms accounted for 66% of the faunal biomass in NT, but only 31% of their biomass in CT where protozoa comprised the largest biomass of soil fauna (57%). In contrast to that of protozoa, earthworm biomass was highest in the cool season, totaling 95 and 36 kg C ha$^{-1}$ in NT and CT, respectively; which is similar to that reported by Hendrix et al. (1987). Notably, their biomass in the warm season was approximately 3 to 6 times lower than that of the cool season in both tillages. As a result ≈79 and 86% of the earthworm respiration could be attributed to the cool season in NT and CT, respectively. Assuming an assimilation efficiency of 0.20 (Persson et al., 1980), their consumption of C totaled 5.4 and 1.9 Mg C ha$^{-1}$ yr$^{-1}$, ≈55 and 19% of the estimated annual C inputs (above- and belowground) to NT (9.8 Mg C ha$^{-1}$ yr$^{-1}$) and CT (9.7 Mg C ha$^{-1}$ yr$^{-1}$), respectively. Parmelee and Crossley (1988) estimated the N-flux from earthworm tissue in NT to be ≈40 kg N ha$^{-1}$ yr$^{-1}$. The biomass of earthworms was much lower at the Kjettslinge and Lovinkhoeve sites. Boström (1988) estimated that the N flux attributable to earthworm excretion and biomass turnover ranged between 3 and 12 kg N ha$^{-1}$ yr$^{-1}$ in the cropping systems at the Kjettslinge site, the lower values being more typical of fertilized barley soils. Though absent from the CF soils, earthworms were estimated to mineralize ≈38 kg N ha$^{-1}$ yr$^{-1}$ in the IF soils at the Lovinkhoeve site (Didden et al., 1994).

The Enchytraeidae were originally hypothesized to play a somewhat greater role in the detrital food web of CT than of NT soil (Hendrix et al., 1986). Subsequently, Parmelee et al. (1990) found that enchytraeid populations and biomass at HSB were higher in NT than in CT soil on at least some sample dates, though the authors acknowledged inefficiencies in their extraction technique. Recent estimates of enchytraeid biomass by van Vliet et al. (1994) are more than an order of magnitude higher than those of Parmelee et al. (1990). While the annual average enchytraeid biomass was significantly higher in NT than in CT, the data of van Vliet et al. (1994) also showed that their biomass was more than twofold higher in the cool-season than in the warm-season soils (Table 2). Enchytraeids are known to be intolerant of the warm, dry conditions that persist throughout much of the summer cropping season at HSB. Results from the Kjettslinge and Lovinkhoeve sites provide an interesting contrast to those of HSB. Although the biomass of all other faunal groups was greater under grass and lucerne leys, the biomass of enchytraeids was highest in the cultivated barely soils (Lagerlöf et al., 1989; Paustian et al., 1990). Similarly, the biomass of enchytraeids was nearly twofold higher in CF than IF soils at the Lovinkhoeve site (Zwart et al., 1994), unlike that of other faunal groups. Based on the revised estimates of their biomass and assuming an assimilation efficiency of 0.25 (50% microbivorous and 50% saprovorous) (Persson et al., 1980), enchytraeids at HSB were estimated to consume ≈230 kg C ha$^{-1}$ yr$^{-1}$ in CT and NT, respectively; with more than two-thirds of that consumption occurring in the cool season. These values compare with an estimated consumption of 180 to 240 kg C ha$^{-1}$ yr$^{-1}$ by enchytraeids under barley at the

Kjettslinge site (Paustian et al., 1990) and 72 to 94 kg C ha$^{-1}$ yr$^{-1}$ under wheat at the Lovinkhoeve site (Didden, 1990b).

Though earthworms and enchytraeids are generally considered detritivores, a significant proportion of their diet can be composed of fungi and fungal byproducts. For example, *Lumbricus terrestris, L. rubellus* and *Apporrectodea caliginosa,* the later two species being dominant at HSB, are known to feed extensively on fungi and fungal-conditioned substrates, probably due to their high protein content (Lee, 1985). Given this fact, it is interesting to note that the F:B ratios in NT were much lower in the cool season when earthworm biomass was nearly fourfold higher than in the warm season. Although earthworm biomass remained much lower in CT, there was no shift in the F:B ratios in spite of seasonal differences in earthworm biomass. Though microorganisms almost certainly contribute significantly to the diet of earthworms and enchytraeids, the relative contributions of bacteria and fungi to the C and N they assimilate remain poorly known.

The selection of specific conversion factors may contribute to errors in the calculations presented above, and these have been discussed previously (Hendrix et al., 1987). For this reason widely cited values were selected in all cases, except where independent estimates were available from HSB data. Furthermore, the metabolic activities and turnover rates of organisms may differ between tillages (Andrén and Lagerlöf, 1983; Golebiowska and Ryszkowski, 1977). Because there are no independent measures of production, respiration, and defecation for organisms in the two tillage systems at HSB, the same values were used in both and thus may tend to de-emphasize the differences between tillages (Hendrix et al., 1987). Furthermore, species-specific differences in these conversion factors may also be important where the composition of the biological communities differs with tillage practice.

Independent estimates of carbon losses from CT and NT soils were made from measurements of crop residue decomposition as a simple validation of the respiratory loss estimates (Table 5). Measured inputs of crop plus weed residues and roots (NT = 9.8 megagram [Mg] C ha$^{-1}$ yr$^{-1}$; CT = 9.7 Mg C ha$^{-1}$ yr$^{-1}$) were used to calculate the decomposition losses of C in each of the cropping seasons using single negative exponential decay rates derived from litterbag studies (Beare et al., 1992; unpublished data). Buried residue decay rates were used to predict the C losses for all inputs in CT. In NT the buried straw decay rates were used to predict root decomposition and surface straw decay rates were used to calculate the losses of C from aboveground residues. The losses of C predicted by the decomposition estimates were remarkably similar to the calculated respiratory losses. Differences between these estimates were slightly greater in NT (±3.7 to 4.9%) than in CT (±0.4 to 1.5%) in both seasons, though both measures predicted lower C losses from NT. As such, these measures of C loss are consistent with the observed differences in SOM standing stocks between NT (30.7 Mg C ha$^{-1}$) and CT (26.1 Mg C ha$^{-1}$). Furthermore, the differences in C losses were much greater in the warm season (781 to 1140 kg C ha$^{-1}$) than the cool season (66 to 230 kg C ha$^{-1}$), which

Table 5    Seasonal Differences in the Calculated Respiratory Losses of C (kg C ha⁻¹)
for Each of the Microbial and Faunal Groups in Conventional Tillage and
No-Tillage Soils at the HSB

| | Summer–autumn[a] | | Winter–spring[a] | | Annual total | |
|---|---|---|---|---|---|---|
| | NT | CT | NT | CT | NT | CT |
| **Fungi** | | | | | | |
| Respiration | 899 | 833 | 800 | 776 | 1,699 | 1,609 |
| % Total | 25.8 | 19.5 | 20.0 | 19.1 | 22.3 | 19.3 |
| **Bacteria** | | | | | | |
| Respiration | 2,253 | 3,159 | 2,448 | 2,904 | 4,701 | 6,063 |
| % Total | 64.7 | 74.1 | 61.2 | 71.4 | 62.8 | 72.8 |
| **Protozoa** | | | | | | |
| Respiration | 132 | 208 | 56 | 104 | 188 | 312 |
| % Total | 3.8 | 4.9 | 1.4 | 2.6 | 2.5 | 3.7 |
| **Nematodes** | | | | | | |
| Respiration | 6.6 | 11.9 | 5.4 | 8.6 | 12.0 | 20.5 |
| % Total | 0.19 | 0.28 | 0.13 | 0.21 | 0.16 | 0.25 |
| **Microarthropods** | | | | | | |
| Respiration | 5.2 | 2.0 | 2.7 | 0.9 | 7.9 | 2.9 |
| % Total | 0.15 | 0.05 | 0.06 | 0.02 | 0.11 | 0.03 |
| **Enchytraeids** | | | | | | |
| Respiration | 12.5 | 10.4 | 26.3 | 23.3 | 38.7 | 33.7 |
| % Total | 0.36 | 0.24 | 0.66 | 0.57 | 0.52 | 0.40 |
| **Earthworms** | | | | | | |
| Respiration | 175 | 41 | 665 | 252 | 840 | 293 |
| % Total | 5.0 | 1.0 | 16.6 | 6.2 | 11.2 | 3.5 |
| Total respiration | 3,483 | 4,264 | 4,003 | 4,069 | 7,486 | 8,334 |
| Decomposition losses of C[b] | 3,160 | 4,300 | 3,720 | 3,950 | 6,980 | 8,250 |

[a] Values are kg C ha⁻¹ season⁻¹ (182 days); summer–autumn — June to November, winter–spring — December to May.
[b] Calculated from measured inputs of crop residues plus roots and estimates of the surface- and buried-residue decomposition rates in each season in NT and CT soils. Units are kg C ha⁻¹ season⁻¹.
Sources of primary data as in Table 3.

corresponds with the greatest differences in microbial, principally bacterial, biomass between the two tillage systems. In NT the increase (15 to 18%) in C loss from the warm season to the cool season is marked by a shift toward a more bacterial-based food web in which the contributions of soil fauna to total soil respiration are nearly doubled.

Seasonal differences in the mineral N content of NT and CT soils may also be attributed to the composition of their belowground food webs. During the warm season at HSB the bacteria-based food web of CT yielded a significantly higher N content than that of the more fungal-dominated food web associated with NT. Although much lower, there were no differences in the mineral N content of CT and NT soils during the cool season when the composition of their microbial communities was relatively more similar. Furthermore, as mentioned previously, the difference in mineral N content of NT and CT soils tends to increase with depth during the warm season, to the extent that the nitrate-enriched pool of mineral N in CT may be more susceptible to leaching below the root zone. This observation is consistent with earlier findings of Stinner et al. (1984), which showed that $NO_3$ leaching losses are considerably higher in CT than NT soils at HSB.

## Statistical Description and Model Simulations

A somewhat more detailed analysis of fungal and bacterial pathways of organic matter processing and nutrient mineralization can be obtained from statistical analyses of population dynamics and model simulations (e.g., Hunt et al., 1987). Moore et al. (1990) constructed food webs for the CF and IF systems of the Lovinkhoeve site using a functional group approach similar to that described above. The authors used canonical discriminant analysis combined with multivariate analysis of variance to distinguish differences in the composition and temporal dynamics of the CF and IF food webs. Their analyses showed that the belowground food webs could be compartmentalized into fungal-, bacterial-, and root-based channels of energy. Furthermore, the temporal dynamics of the principal functional groups differed significantly in IF, but not in CF. Bacteria and fungi exhibited different temporal dynamics in IF, as did their consumers in the bacterial and fungal energy channels. The temporal dynamics of the root energy channel also differed from that of fungi, bacteria, and bacterivorous fauna.

Moore and de Ruiter (1991) also showed how model simulations of N fluxes through fungal and bacterial channels could be used to illustrate differences in the vertical stratification of N dynamics in CF and IF systems. Whereas the total N flux rate (kg N ha$^{-1}$ 10 cm$^{-1}$ yr$^{-1}$) did not differ with depth in CF, the total N flux in the top 10 cm of IF was more than double that of the 10- to 25-cm depth. Furthermore, more of the vertical stratification in N flux rate could be attributed to the consumers of fungi and bacteria than to bacteria and fungi themselves (Table 6). Nonetheless, ≈97 and 99% of the total N flux could be attributed to the bacterial pathway in IF and CF, respectively. Similar models have been used to predict the contributions of microbivorous and predatory fauna to N mineralization. For example, De Ruiter et al. (1993) showed that, in spite of their relatively low biomass, bacterial-feeding and predatory nematodes each contribute (both directly and indirectly) on the order of 8 to 19% of the N mineralized in the CF and IF soils at the Lovinkhoeve site.

As is apparent from the above discussion, the development of alternative management strategies to achieve greater sustainability of the soil resource will require an understanding of how soil biota respond both spatially and temporally to changes in the quantity, timing, and placement of organic residues. This conclusion is likely to apply equally well to other exogenous inputs such as animal manures, mineral fertilizers, and pesticides.

Table 6   Estimates of N Flux (kg N ha$^{-1}$ 10 cm$^{-1}$ yr$^{-1}$) Through Bacteria and Fungi and the N Passing Through Microbial Consumers and Predators in Bacterial and Fungal Energy Channels

| | Biomass N | | Energy channel | |
|---|---|---|---|---|
| Dept (cm) | CF | IF | CF | IF |
| **Bacteria** | | | | |
| 0–10 | 32 | 49 | 30 | 55 |
| 10–25 | 36 | 40 | 30 | 25 |
| **Fungi** | | | | |
| 0–10 | 0.60 | 1.10 | 0.70 | 0.85 |
| 10–25 | 0.70 | 0.90 | 0.70 | 0.55 |

From Moore, J. C. and de Ruiter, P. C., 1991. *Agric. Ecosystems Environ.*, 34:371–397. © 1991 with kind permission of Elsevier Science, 1055 KV Amsterdam, The Netherlands.

## RESIDUE DECOMPOSITION

The effective management of crop residues is recognized as an important aspect of low-input sustainable crop production systems. The importance of residue quality (e.g., nutrient content, C:N ratio, and lignin:N ratio) to determining rates of residue decay and nutrient release is well known (Swift et al., 1979). Where residue quality is constant, the microclimatic conditions imposed on residues are probably primarily responsible for regulating these processes. In arable soils the positioning of organic matter within the soil profile depends largely on the allocation of C to roots and shoots and the method of seed bed preparation used (e.g., moldboard ploughing, chisel ploughing, no-tillage). Under no-tillage (NT) management, crop residues accumulate on the soil surface as a mulch, whereas, with conventional tillage (CT) practices, ploughing results in the fragmentation and burial of crop residues. As such, placement determines the microclimatic conditions of residues (Blevins et al., 1984) and their proximity to exogenous nutrients (Christensen, 1986), factors that influence the structure and function of detrital food webs (Doran, 1980; Hendrix et al., 1986; Mueller et al., 1990; Beare et al., 1993). These in turn determine rates of residue decomposition and patterns of nutrient release (Holland and Coleman, 1987; Beare et al., 1992).

### Residue-Borne Microbial and Faunal Communities

Many studies have described the succession of organisms colonizing plant residues (e.g., Harper and Lynch, 1985; Struwe and Kjøller, 1985; Ponge, 1991; Beare et al., 1993). It is clear from these that the chemical composition of crop residues is an important determinant of both the size and composition of microbial (Broder and Wagner, 1988) and faunal (Parmelee et al., 1989; Beare et al., 1989) communities.

Results of studies at HSB also indicate that residue placement, more so than tillage, has an overriding influence on the size and composition of decomposer communities (e.g., Mueller et al., 1990; Beare et al., 1992). Not surprising perhaps, microbial and faunal populations can be several times greater on buried residues than surface residues, regardless of tillage (Beare et al., 1992), though this seems to depend greatly on climatic conditions. Under drought conditions, the densities and biomass of microbes and fauna on decaying rye straw (*Secale cereale* L.) were found to be much lower on NT-surface residues than CT-buried residues (Beare et al., 1992). However, under normal climatic conditions for Georgia, direct counts of total and FDA-active fungi, total bacteria and ratios of fungal-to-bacterial substrate-induced respiration (SIR) indicated that fungi comprise a larger proportion of the active residue-borne microbial community on NT-surface residues, while bacteria and fungi share more equal importance on CT-buried residues (Table 7) (Beare and Coleman, 1994). Although ratios of fungal-to-bacterial SIR on incorporated residues (CT) tend to remain relatively constant throughout their decay (Beare, unpublished), Neely et al. (1991) have shown that fungal contributions to total SIR on a wide range of surface-applied (NT) residues tend to decrease with increases in residue decay, indicating a shift from a fungal-dominated microbial community in the early stages of decay to a more bacterial-dominated community in the later stages of decay. The initial lignin:N ratio of surface-applied residues was found to explain most of the variation in fungal (73%) and bacterial (59%) SIR.

**Table 7  Effects of Fungicide and Tillage Treatments on Fungal and Bacterial Populations, Ratios of Fungal-to-Bacterial SIR, Residue Decay Rates, and the Percentage of Cellulose and Lignin Remaining after 45 Weeks of *Secale* Residue Decomposition**

| Measured parameter | No-tillage | | Conventional tillage | | Tillage |
| --- | --- | --- | --- | --- | --- |
| | Control | Fungicide | Control | Fungicide | ($p$ <0.05) |
| Total fungi (m/g AFDW)[a] | 2,147 | 698* | 1,806 | 747* | NT > CT |
| FDA fungi (m/g AFDW)[a] | 89.8 | 16.4* | 55.5 | 21.4* | NT > CT |
| Total bacteria ($10^9$/g AFDW)[a] | 12.4 | 12.0 | 14.9 | 16.3 | CT > NT |
| Fungal-to-bacterial SIR ratios[a] | 1.63 | 0.65* | 1.01 | 0.48* | NT > CT |
| Decay rate (k $yr^{-1}$) | 1.35 | 1.06* | 1.93 | 1.76* | CT > NT |
| Cellulose (% remaining)[b] | 41.1 | 58.9* | 29.6 | 32.4 | NT > CT |
| Lignin (% remaining)[b] | 80.9 | 92.0* | 79.0 | 88.5* | NS |

*Note:* AFDW = ash-free dry weight. Asterisks indicate significant differences ($p$ <0.05) between the fungicide and control treatments within tillage practice.

[a] Values are averages of four sample dates.
[b] Values are the percentage of the initial amounts remaining after 45 weeks of residue decay.

Adapted from Beare, M. H. and Coleman, D. C., 1994.

Residue management is also an important determinant of fungal community composition. For example, recent studies at HSB (Beare et al., 1993) indicate a strong differentiation of the fungal community into surface residue specialist (e.g., *Alternaria alternata, Epicoccum nigrum, Phoma* spp., etc.) and soil specialist (e.g., *Aspergillus* spp., *Trichoderma viridis, Penicillium verruculosum,* etc.), while buried residues contain elements of both surface residue and soil specialist communities. These findings are consistent with those reported elsewhere (Harper and Lynch, 1985; Broder and Wagner, 1988). The history of residue management at a given site can also influence microbial community composition and activity. Killham et al. (1988) have shown that repeated incorporation of barley straw can markedly increase populations of soil fungi, particularly cellulolytic fungi, affecting a significant increase in the rate of straw decomposition.

Although populations of residue-borne micro- and mesofauna are often much lower on surface applied (NT) as compared with incorporated residues (CT), there are notable differences in the composition of these faunal communities as well. For example, Mueller et al. (1990) showed that Oribatida (46%) and Prostigmata (45%) dominate the mite community on surface-applied *Secale* residues in NT, while Oribatida (47%) and Mesostigmata (36%) are most common on CT-buried residues. However, the significance of these observations from a functional standpoint is much less clear. Despite marked effects of both tillage practice and residue placement on the composition of mite communities, in all cases ≈85 to 90% of the mites were identified as fungivores. In contrast, Beare et al. (1992) showed that, during periods of normal climatic conditions, ratios of fungivore-to-bacterivore biomass were two to three times higher on surface residues of NT (3.7 to 4.7) than buried residues of CT (1.9 to 2.2). Furthermore, in a related study of weed residue decomposition in NT soils of Georgia, Parmelee et al. (1989) found that the composition of residue-borne microbivorous nematode and microarthropod communities was affected as much by site as by the quality of the residues. They argued that surface-applied crop residues could function as resource "islands" in soils of lower organic matter content, leading to greater colonization by microbivorous fauna.

## Controls on Residue Decomposition and N Mineralization

Several studies indicate that measures of the size, composition, and activity of residue-borne microbial communities can be useful predictors of residue decay rates (Widden et al., 1986; Robinson et al., 1993). Neely et al. (1991), for example, showed that the respiratory response of the residue-borne microbial community to the addition of a labile substrate such a glucose (as measured by SIR) can provide a valuable index of residue decomposition rates. Distinguishing the relative contributions of fungi and bacteria has proved to be somewhat more difficult.

The importance of residue placement and the resulting decomposer food webs to determining rates of residue decomposition and nutrient release are illustrated by several studies at HSB (Mueller et al., 1990; Beare et al., 1992; Beare and Coleman, 1994). Consistent with their higher populations of decomposers, the decomposition rate of incorporated residues in CT ranged from 1.4 to 1.9 times greater than that of surface residues in NT (Table 7) (Figure 2). Based on results of SIR assays, fungi appear to play a somewhat greater role than bacteria in the breakdown of simple carbonaceous substrates on NT-surface residues, while bacteria play a somewhat greater role on CT-buried residues (Beare and Coleman, 1994). Notably, the overall mean rates of total SIR measured in these studies did not differ significantly between NT-surface (525 μg $CO_2$-C/g AFDW/h) and CT-buried (503 μg $CO_2$-C $g^{-1}$ AFDW $h^{-1}$) residues. These results suggest that the biomass and activity potentials of residue-borne microbial communities are similar in NT and CT, despite a much higher decay rate for CT-buried residues (1.93 $yr^{-1}$) than for NT-surface residues (1.35 $yr^{-1}$). Though these findings appear inconsistent, fluctuations in the microclimatic conditions of surface residues impose a greater periodicity on microbial growth and activity, which is expected to contribute to slower rates of residue decay in NT as compared with CT.

In other studies at HSB, biocides were used to inhibit fungi, bacteria, and microarthropods in the field as a means to quantify their role in residue decomposition and nutrient mineralization in CT and NT soils (Mueller et al., 1990; Beare et al., 1992, 1993; Beare and Coleman, 1994). For example, experiments with the fungicide captan have shown that where populations of total and FDA-active fungi are reduced by 50 to 70% in both tillages, the decomposition rates of NT surface residues are reduced by 21 to 36%, almost twice that of CT-buried residues (9 to 21%) (Table 7) (Figure 2C and G). Furthermore, the contributions of fungi to the decomposition of surface residues in NT involves primarily the cellulose and lignin constituents of the residues, while in CT, the reduction in decomposition appears to involve only the lignin component, affecting little or no change in the breakdown of cellulose or cell-soluble constituents (Table 7). This result may be attributed to the differences in fungal community composition noted above. Effects of the bactericide treatment were somewhat different. In general, the bactericide inhibited the decay of CT-buried residues (35%) much more than that of NT-surface residues (25%), despite similar reductions in bacterial populations (Beare et al., 1992). Overall, these studies indicate that fungi play a greater role in regulating the decomposition of surface-applied residues such as occur under minimum or NT management, while bacteria contribute relatively more to the decay of incorporated residues in cultivated soils.

The activities of bacteria, fungi, and microbivorous fauna determine the balance between the mineralization and immobilization of nutrients and, consequently, the release of nutrients to the mineral soil. As such, crop residues act as both a source and sink of nutrients. In the fungicide experiments discussed above and inhibition of fungi markedly altered the fluxes of N from

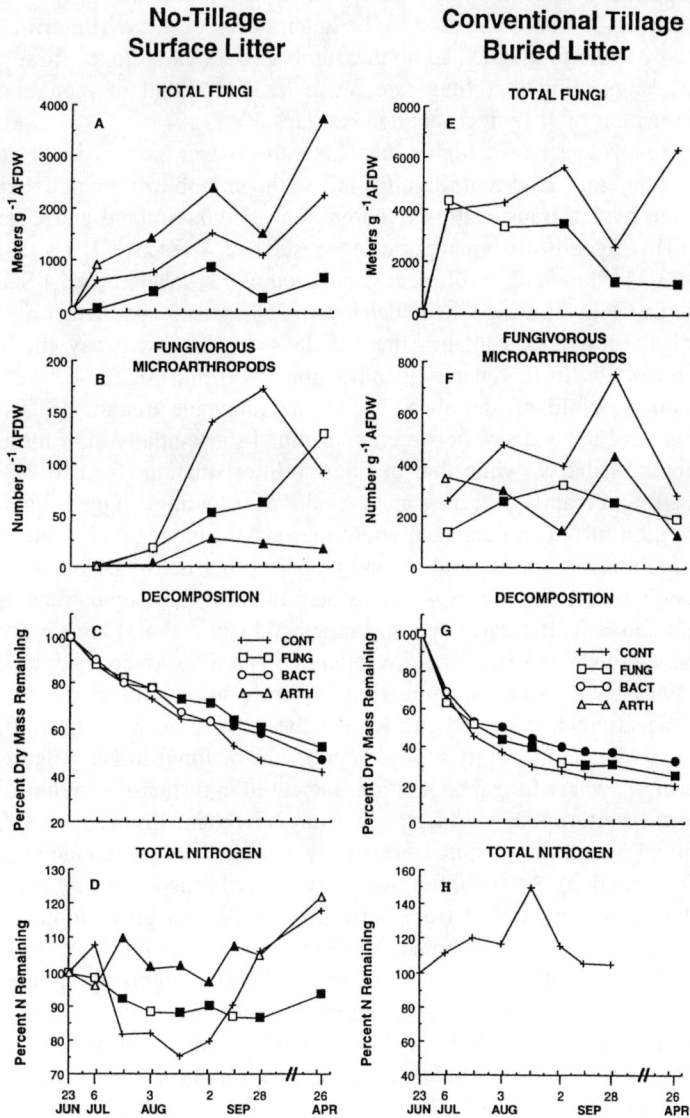

**Figure 2** Effects of biocide treatments (*CONT* = Control, *FUNG* = Fungicide, *BACT* = Bactericide, *ARTH* = Arthropod exclusion) on populations of fungi (**A,E**) and fungivorous microarthropods (**B,F**) and the percentage of dry mass (**C,G**) and nitrogen (**D,H**) remaining during the decay of surface-applied and buried residues in NT and CT agroecosystems, respectively. Effects of the biocides are shown only where the treatments differed significantly (ANOVA/Tukey-Kramer) from controls across dates. Solid symbols indicate a significant difference (ANOVA/Tukey-Kramer) from the control on each sample date. (Adapted from Beare et al., 1992.)

surface residues in NT, slowing net N mineralization in the early stages of decay and net N immobilization in the latter stages of decay (Figure 2D). In CT, however, the fungicide had no measurable effect on N fluxes from buried residues, suggesting that fungi are much less important in regulating the immobilization of N by incorporated residues.

Where residues have a high initial C:N ratio, as was the case in the studies at HSB, fungi may contribute significantly to the immobilization of exogenous N through hyphal translocation (Andrén et al., 1990; Holland and Coleman, 1987). This hypothesis is supported by results of a recent $^{15}$N tracer study (Figure 3). In this field experiment, a solution of $^{15}$N-labeled $(NH_4)_2SO_4$ was injected just below ($\approx$2 cm) the mulch layer of NT soils treated with or without (control) the fungicide captan. After 128 days of rye straw decay, the immobilization of $^{15}$N in the coarse-litter fraction (>2.0 mm) of the control plots was about fourfold greater than that of the fungicide treatments. The $^{15}$N enrichment of the coarse-litter fraction declined substantially after more than 10 months of decay, while that of the fine-litter fraction (0.25 to 2.0 mm) increased significantly over the same period. These findings suggest that fungal translocation of N can be an important mechanism for regulating the uptake and immobilization of mineral N (and perhaps other nutrients) in NT soils.

Results of the biocide experiments also indicate that populations of fungivorous fauna (particularly Oribatid mites and Collembola) are tightly coupled to the growth and activity of mycelial fungi on NT-surface residues (Beare et al., 1992). Whereas populations of fungivorous microarthropods were slow to colonize fungicide-treated surface residues, the removal of these fungal-feeders yielded significantly higher populations of fungi in NT (Figure 2A). Furthermore, where fungal populations increased in response to reduced grazing by microarthropods, surface residues in NT retained more than 100% of their initial N content, in spite of relatively small effects on residue mass loss (Figure 2C and D). As such, fungivorous microarthropods are probably more important in mobilizing N from surface residues through their grazing on fungal than in contributing to residue decomposition. Therefore the processes of fungal translocation and assimilation and fungivore grazing appear to be the most important factors regulating the amount and timing of N releases from surface-applied residues. The population dynamics of protozoa were also found to track the changes in bacterial populations on surface residues (Beare et al., 1992). In CT, however, the population dynamics of microbivorous fauna were decoupled from microbes, and there were no measurable effects of the biocides on litter N fluxes (Figure 2E to H). From these studies Beare et al. (1992) estimated that the interactions between fungi and fungal-feeding microarthropods may be responsible for up to 60% of the net N losses from surface-applied residues in the early stages of decay, while fungi account for as much as 86% of the N immobilized by high C:N ratio residues in the later stages of decay.

Support for these observations can be found in the studies of Andrén (1987) and Andrén and Paustian (1987) in the Swedish Arable Land Project. Their

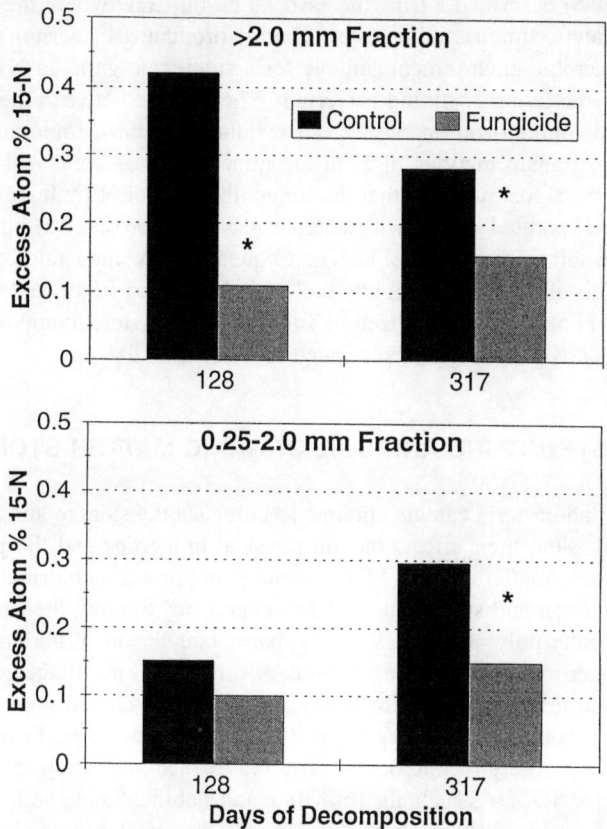

**Figure 3**   Effects of fungicide applications on the immobilization of mineral soil associated $^{15}N$-labeled $NH_4$ by the >2.0 mm and 0.25 to 2.0 mm fractions of surface
residues in NT soils. (From Beare, M. H. and Coleman, D. C., unpublished.)

results showed that differences in the immobilization of N by high C:N ratio
barley straw depended on the time of residue return, the cropping practice,
and residue placement. Furthermore, the differences they reported seem to be
related primarily to mineral N availability. Buried straw immobilized less N
than surface-applied straw in soils cropped to fescue or lucerne, probably due
to competition for available N between plant roots and residue-borne decomposers. The immobilization of N coincided with the initial ingrowth of fungal
mycelium into the decaying barley straw (Wessen and Berg, 1986). Fungal
contributions to N immobilization were estimated to be approximately three
and a half times greater than for bacteria, though much of the immobilized N
was estimated to be in microbial byproducts rather than in the microbial
biomass.

The interactions between microorganisms colonizing straw can also have
important consequences for their nutrient dynamics. For example, Lynch and

Harper (1985) described a tripartite association on straw where the activities of a cellulolytic fungus and a polysaccharide-producing bacterium yielded a C-rich, anaerobic environment suitable for a significant gains in N (84 kg N ha$^{-1}$) by a free-living N$_2$-fixing bacterium. The balance between the mineralization and immobilization of N from residues can have important consequences for transformations of N in the mineral soil as well. Aulakh et al. (1991) showed, for example, that the immobilization of N by high C:N ratio residues incorporated in soils with a high water-filled porosity can significantly deplete the soil mineral N pool and, consequently, slow their rates of denitrification. This effect may be slightly greater in NT where significant quantities of mineral N may be immobilized by fungi in the slow-to-decompose surface residues as compared with incorporated residues of CT.

## SOIL STRUCTURE AND SOIL ORGANIC MATTER STORAGE

Fungi and bacteria can also indirectly influence the storage and release of nutrients through their effects on soil physical properties and the protection of soil organic matter (SOM). Mycelial fungi and bacteria contribute directly to the formation and stabilization of soil aggregates through their deposition of extracellular polysaccharides and hyphal entanglement. Where aggregates remain intact for relatively long periods of time (as in no-tillage soils), clay surfaces and micropores (<1 μm diameter) become occluded with the extracellular byproducts of microorganisms (Adu and Oades, 1978; Foster, 1981), restricting the access of microorganisms and extracellular enzymes to physically isolated SOM. As such, the formation and stabilization of soil aggregates represent a potentially important mechanisms for the storage and protection of SOM (Edwards and Bremner, 1967; Elliott, 1986; Gupta and Germida, 1988).

The contributions of fungi and bacteria to soil aggregation have been evaluated in several laboratory studies. Aspiras et al. (1971) showed that polysaccharides and humic substances were the dominant binding agents produced by several species of bacteria and streptomyces. In contrast, humic and lignin-like compounds were found to be the principal binding agents produced by most species of fungi. Their studies also suggested that aggregates bound by filamentous structures (and/or their associated binding agents) are more resistant to physical disruption than those bound by bacterial polysaccharides. The stabilization of macroaggregates (>250 μm) is most often attributed to fungi, while that of microaggregates is usually associated with the production of adhesive metabolites by bacteria. Furthermore, it is generally accepted that fungal polysaccharides are more stable to degradation than those derived from bacteria (Burns and Davies, 1986).

As discussed previously, fungal-based food webs often support large populations of microarthropods (especially mites and Collembola) and earthworms. In some soils, especially those of temperate regions, the feeding activities of litter-dwelling microarthropods may be responsible for significant accumulations of fecal pellets in the surface soil (Rusek, 1985). Earthworms also influence the stabilization of aggregates both directly, through the rearrangement of particles and the deposition of mucus, and indirectly, through the stimulation of microbial, especially fungal activity (Marinissen and Dexter, 1990). With proper conditioning the fecal pellets and casts of these organisms form stable aggregates that contain their excretory byproducts, as well as fragmented plant and microbial debris. Once stabilized, these fecal aggregates can function as anaerobic microsites where conditions are suitable for high levels of denitrification (Elliot et al., 1990). In bacterial-based food webs, the feeding of protozoa on bacteria in the micropores of aggregates (Elliott et al., 1980; Foster and Dormaar, 1991) may contribute to increased organic matter turnover and reduce aggregate stability.

The changes in soil structure that result from intensive cultivation are often associated with significant losses of SOM (Tisdall and Oades, 1982; Elliott, 1986). The magnitude of these effects depends greatly on the intensity of cultivation and the quantity and quality of fertilizers and organic residues returned to the soil (Rasmussen and Collins, 1991). Minimum and no-tillage (NT) soils often support higher standing stocks of SOM and greater soil aggregation than conventionally tilled (CT) soils (Bruce et al., 1990; Havlin et al., 1990; Carter, 1992). Results of recent studies at HSB (Beare et al., 1994a,b) suggest that the formation and stabilization of macroaggregates in NT soils represent important mechanisms for the protection and maintenance of SOM that may otherwise be lost under CT practices. This mechanism may explain the nearly 18% higher standing stock of soil organic matter in NT (30.7 Mg C ha$^{-1}$) as compared with CT (26.1 Mg C ha$^{-1}$) soils at this site. In the studies mentioned above, Beare et al. (1994a,b) found that macroaggregate-protected pools of SOM accounted for 18.8% of the total mineralizable C (914 kg ha$^{-1}$) in NT (0 to 15 cm), but only 10.2% of the total mineralizable C (832 kg ha$^{-1}$) in CT. Furthermore, nearly all of the difference between tillages in aggregate protected C was found in the surface soils where the greatest differences in microbial community composition have been identified.

Many of the effects of CT and NT management on C and N dynamics may be attributed to differences in the composition of their decomposer communities. As discussed previously, mycelial fungi appear to play a primary role in regulating the decomposition of surface-applied crop residues in NT, while bacteria are more important to the decomposition of incorporated residues in CT (Hendrix et al., 1986; Beare et al., 1992). Recent studies (Beare et al., 1994c) indicate that mycelial fungi also contribute more significantly to the formation and stabilization of soil aggregates in NT than in CT soils. These conclusions are based on studies using the fungicide Captan to manipulate fungi in fields. In the later experiment, a 42% reduction in fungal hyphae in the fungicide-treated soils of NT resulted in a 40% decline in the largest macroaggregates and redistribution of particles into smaller size classes (Table 8). Despite similar reductions in hyphal densities, there were no significant

**Table 8    Effects of Fungicide on Water-Stable Aggregate Distributions in Surface Soils (0–5 cm) of Conventional and No-Tillage Plots at HSB**

| Aggregate size-class | % Water-stable aggregates[a,b] | | | | Tillage (p <0.05) |
|---|---|---|---|---|---|
| | No-tillage | | Conventional tillage | | |
| | Control | Fungicide | Control | Fungicide | |
| >2,000 µm | 65.5 | 42.3* | 43.2 | 38.7 | ** |
| 250–2,000 µm | 17.2 | 31.2* | 20.1 | 23.8 | |
| 106–250 µm | 5.9 | 10.0* | 9.9 | 10.4 | ** |
| 53–106 µm | 2.5 | 5.8* | 3.5 | 2.9 | |
| <53 µm | 8.9 | 10.6 | 23.4 | 24.2 | ** |

[a] Values are normalized to a sand-free basis.
[b] Asterisks indicate significant (ANOVA/LSD; $p < 0.05$) effects of the fungicide within tillage and aggregate size-class (*) and the main effects of tillage within size-class (**).
Adapted from Beare, M. H., et al., in press.

effects of the fungicide on WSA distributions in CT. Furthermore, the greater abundance of macroaggregates in surface soils (0 to 5 cm) of NT corresponds with a 1.3 to 1.5 times higher density of fungal hyphae, a nearly twofold higher concentration of acid-hydrolyzable carbohydrates and a greater contribution of microbial- than plant-derived sugars as compared with CT soils. These findings are supported by those of Gupta and Germida (1988), who found much greater macroaggregation and higher populations of fungi in native sod as compared with long-term (69 yr) cultivated soils. Furthermore, the rapid decline in aggregate stability often noted in bare fallow soils or in soils where organic matter inputs are removed by burning may be attributed as much to hyphal disintegration as to reductions in polysaccharide binding agents (Tisdall and Oades, 1982).

Differences in composition of soil biotic communities that result from different soil management practices are likely to be important to determining the size and quality of aggregate protected and unprotected pools of SOM. For this reason, a knowledge of the composition of soil biotic communities under different cropping systems may be useful in predicting changes in soil structure and the accumulation and loss of SOM.

## CONCLUSIONS

Understanding how agricultural management practices contribute to the sustained fertility and productivity of arable soils requires a knowledge of both the structure and function of belowground food webs with particular attention to their spatial and temporal variances. Results of the long-term tillage trial at HSB, as well as those of other arable lands programs, have suggested some important ways that cultivation of soils can affect a change in the structure and function of belowground food webs. They have also helped to identify some of the key mechanisms by which fungal- and bacterial-based food webs regulate soil processes. For example, where soils are disturbed by ploughing and the crop residues incorporated, there is much less spatial and temporal differentiation of food web structure and function. Under CT practices, bacterial-based food webs play a relatively greater role in regulating SOM dynamics and nutrient cycling. Consistent with their higher rates of decay, incorporated residues retain larger populations of both primary and secondary decomposers than do surface applied residues, yet the population dynamics of microbivorous fauna in CT appear to be much less tightly coupled to that of the microbial populations. In CT, the soil microbial biomass also makes up a much higher proportion of the whole soil C than in NT. Furthermore, a greater proportion of this biomass is bacterial in origin and concentrated in the plough layer where it is susceptible to the flushes of mineralization imposed by tillage events and wet/dry cycles. Overall, these properties of cultivated soils appear to contribute to greater organic matter losses and lower nutrient retention.

In contrast, where cultivation is minimized and crop residues are retained on the soil surface, there is a much greater spatial and temporal differentiation of belowground food webs and the processes that they mediate as compared with conventionally cultivated soils. In NT systems, for example, fungal-based food webs often develop on surface-applied crop residues. In this environment, mycelial fungi play a key role in the immobilization of mineral N; however, it is their interactions with fungal-feeding fauna that determine the amount and timing of N releases from decaying residues. The vertical stratification of belowground food webs is also much more pronounced under NT, shifting from a fungal-based food web near the soil surface to a more bacterial-based food web deeper in the profile. Where fungi are concentrated near the mineral soil surface, they play an important role in the stabilization of aggregates formed by the casting activities of meso- and macrofauna (e.g., microarthropods, earthworms). This and other mechanisms of aggregate stabilization can significantly enhance the physical protection of organic matter, yielding greater SOM storage in NT soils. The microbial and faunal communities of NT are also marked by a much greater seasonal dependence than is found in CT. In NT, the shift from a bacterial-dominated biomass in the cool season to a more fungal-dominated biomass in the warm season corresponds with a significant decrease in C losses relative to CT where bacterial-based food webs persist throughout the year. The shift toward a more fungal-based food web in NT also appears to contribute to a lower mineralization and greater retention of N as compared with the bacterial-based food web in CT. Furthermore, the greater stratification of mineral N near the soil surface in NT helps to minimize losses of N by leaching, though denitrification losses of N are greater in surface soils where the availability of C and anaerobic microsites is less limiting.

Determining the extent to which differences in the structure and function of soil food webs can be used to predict changes in the sustainability of agricultural management practices remains an important area for future research.

## ACKNOWLEDGMENTS

The author is grateful for the encouragement and support of numerous colleagues at the University of Georgia: M. L. Cabrera, D. C. Coleman, D. A. Crossley, Jr., W. L. Hargrove, P. F. Hendrix, S. Hu, C. L. Neely, B. R. Pohlad, D. H. Wright, and P. J. van Vliet. The research findings reported here for the HSB site were supported by NSF grants (BSR-8506374 and BSR-8818302) to the University of Georgia Research Foundation. Support for the preparation of this chapter was provided by the Institute for Crop and Food Research, Christchurch, New Zealand.

## REFERENCES

Adu, J. K. and Oades, J. M., 1978. Utilization of organic materials in soil aggregates by bacteria and fungi. *Soil Biol. Biochem.,* 10:117–122.

Anderson, J. M., 1988. Spatiotemporal effects of invertebrates on soil processes. *Biol. Fertil. Soils,* 6:216–227.

Anderson, J. P. E. and Domsch, K. H., 1973. Quantification of bacterial and fungal contributions to soil respiration. *Archiv für Mikrobiologie,* 93:113–127.

Andrén, O., 1987. Decomposition of shoot and root litter of barley, lucerne, and meadow fescue under field conditions. *Swed. J. Agric. Res.,* 17:113–122.

Andrén, O. and Lagerlöf, J., 1983. Soil fauna Microarthropods, Enchytraeids and Nematodes in Swedish agricultural cropping systems. *Acta Agriculturae Scandinavica,* 35:33–52.

Andrén, O., Lindberg, T., Paustian, K., and Rosswall, T., 1990. *Ecology of Arable Lands — Organisms, Carbon and Nitrogen Cycling.* Ecological Bulletins, Copenhagen, Ecological Bulletin series #40, 222 pp.

Andrén, O. and Paustian, K., 1987. Barley straw decomposition in the field: a comparison of models. *Ecology,* 68:1190–1200.

Aspiras, R. B., Allen, O. N., Harris, R. F., and Chester, G., 1971. The role of microorganisms in the stabilization of soil aggregates. *Soil Biol. Biochem.,* 3:347–353.

Aulakh, M. S., Doran, J. W., Walters, D. T., Mosier, A. R., and Francis, D. D., 1991. Crop residue type and placement effects on denitrification and mineralization. *Soil Sci. Soc. Am. J.,* 55:1020–1025.

Aulakh, M. S., Rennie, D. A., and Paul, E. A., 1984. Gaseous nitrogen losses under zero-till as compared to conventional-till management systems. *J. Environ. Qual.,* 13:130–136.

Babiuk, L. A. and Paul, E. A., 1970. The use of fluorescein isothiocyanate in the determination of the bacterial biomass of a grassland soil. *Can. J. Microbiol.,* 16:57–62.

Bakken, L. and Olson, R. A., 1983. Bouyant densities and dry matter contents of microorganisms: conversion of a measured biovolume into biomass. *Appl. Environ. Microbiol.,* 45:1188–1195.

Beare, M. H., Blair, J. M., and Parmelee, R. W., 1989. Resource quality and trophic responses to simulated throughfall: effects on decomposition and nutrient flux in a no-tillage agroecosystem. *Soil Biol. Biochem.,* 21:1027–1036.

Beare, M. H. and Coleman, D. C., 1994. Fungal regulation of *Secale* residue decomposition and nitrogen release in conventionally plowed (CT) and no-tillage (NT) soils of the southeastern USA, in *Soil Biota: Management in Sustainable Farming Systems,* Pankhurst, C. E., Ed., CSIRO, Australia, pp. 53–55.

Beare, M. H., Cabrera, M. L., Hendrix, P. F., and Coleman, D. C., 1994a. Aggregate-protected and unprotected pools of organic matter in conventional and no-tillage soils. *Soil Sci. Soc. Am. J.,* 58:787–795.

Beare, M. H., Hendrix, P. F., and Coleman, D. C., 1994b. Water-stable aggregates and organic matter fractions in conventional and no-tillage soils. *Soil Sci. Soc. Am. J.,* 58:777–786.

Beare, M. H., Hu, S., Coleman, D. C., and Hendrix, P. F., in press. Influences of mycelial fungi on soil aggregation and soil organic matter retention in conventional and no-tillage soils. *Appl. Soil Ecol.*

Beare, M. H., Neely, C. L., Coleman, D. C., and Hargrove, W. L., 1990. A substrate-induced respiration (SIR) method for measurement of fungal and bacterial biomass on plant residues. *Soil Biol. Biochem.,* 22:585–594.

Beare, M. H., Neely, C. L., Coleman, D. C., and Hargrove, W. L., 1991. Characterization of a substrate-induced respiration method for measuring fungal, bacterial and total microbial biomass on plant residues. *Agric. Ecosys. Environ.,* 34:65–73.

Beare, M. H., Parmelee, R. W., Hendrix, P. F., Cheng, W., Coleman, D. C., and Crossley, D. A., 1992. Microbial and faunal interactions and effects on litter nitrogen and decomposition in agroecosystems. *Ecol. Monogr.,* 62:569–591.

Beare, M. H., Pohlad, B. R., Wright, D. H., and Coleman, D. C., 1993. Residue placement and fungicide effects on fungal communities in conventional and no-tillage soils. *Soil Sci. Soc. Am. J.,* 57:392–399.

Blevins, R. L., Smith, M. S., and Thomas, G. W., 1984. Changes in soil properties under no-tillage, in *No-Tillage Agriculture: Principles and Practices,* Phillips, R. E. and Phillips, S. H., Eds., Van Nostrand Reinhold, New York, pp. 190–230.

Bloem, J., Lebbink, G., Zwart, K. B., Bouwman, L. A., Burgers, S. L. G. E., de Vos, J. A., and de Ruiter, P. C., 1994. Dynamics of microorganisms, microbivores and nitrogen mineralization in winter wheat fields under conventional and integrated management. *Agric. Ecosys. Environ.,* 51:129–143.

Boström, U., 1988. The Ecology of Earthworms in Arable Land. Population Dynamics and Activity in Four Cropping Systems, Ph.D. thesis, Dept. of Ecology and Environmental Research, Report 34. Swedish Univ. of Agricultural Sciences, Uppsala, p. 147.

Broder, M. W. and Wagner, G. H., 1988. Microbial colonization and decomposition of corn, wheat and soybean residues. *Soil Sci. Soc. Am. J.,* 52:112–117.

Bruce, R. R., Langdale, G. W., and Dillard, A. L., 1990. Tillage and crop rotation effect on characteristics of a sandy surface soil. *Soil Sci. Soc. Am. J.,* 54:1744–1747.

Brussaard, L., Bouwman, L. A., Geurs, M., Hassink, J., and Zwart, K. B., 1990. Biomass, composition and temporal dynamics of soil organisms of a silt loam soil under conventional and integrated management. *Neth. J. Agric. Sci.,* 38:283–302.

Brussaard, L., van Veen, J., Kooistra, M. J., and Lebbink, G., 1988. The Dutch programme on soil ecology of arable farming systems. I. Objectives, approach and some preliminary results. *Ecol. Bull.,* 39:35–40.

Burns, R. G. and Davies, J. A., 1986. The microbiology of soil structure, in *The Role of Microorganisms in a Sustainable Agriculture,* Lopez-Real, J. M. and Hodges, R. D., Eds., A.B. Academic Publishers, Berkhamstead, pp. 9–27.

Carter, M. R., 1992. Influence of reduced tillage systems on organic matter, microbial biomass, macro-aggregate distribution and structural stability of the surface soil in a humid climate. *Soil Till. Res.,* 23:361–372.

Christensen, B. T., 1986. Barley straw decomposition under field conditions: effect of placement and initial nitrogen content on weight loss and nitrogen dynamics. *Soil Biol. Biochem.,* 18:523–529.

Clarholm, M., 1981. Protozoan grazing of bacteria in soil — impact and importance. *Microb. Ecol.,* 7:343–350.

Clarholm, M., 1985. Possible roles of roots, bacteria, protozoa and fungi in supplying nitrogen to plants, in *Ecological Interactions in Soil: Plants, Microbes and Animals,* Fitter, A. H., Atkinson, D., Read, D. J., and Usher, M. B., Eds., British Ecological Society Special Publication No. 4, Blackwell Scientific Publications, Oxford, UK, pp. 355–366.

Coleman, D. C., 1985. Through a ped darkly: an ecological assessment of root-soil-microbial-faunal interactions, in *Ecological Interactions in Soil,* Fitter, A. H., Atkinson, D., Read, D. J., and Usher, M. B., Eds., Blackwell, Oxford, pp. 297–317.

Coleman, D. C., Anderson, R. V., Cole, C. V., McClellan, J. F., Woods, L. E., Trofymow, J. A., and Elliott, E. T., 1984. Roles of protozoa and nematodes in nutrient cycling, in *Microbial-Plant Interactions,* Todd, R. L., Ed., ASA Publication no. 47, Madison, WI, pp. 17–28.

de Ruiter, P. C., Moore, J. C., Zwart, K. B., Bouwman, L. A., Hassink, J., Bloem, J., De Vos, J. A., Marinissen, J. C. Y., Didden, W. A. M., Lebbink, G., and Brussaard, L., 1993. Simulation of nitrogen mineralization in the below-ground food webs of two winter wheat fields. *J. Appl. Ecol.,* 30:95–106.

Didden, W. A. M., 1990a. Involvement of Enchytraeidae (Oligochaeta) in soil structure evolution in agricultural fields. *Biol. Fertil. Soils,* 9:152–158.

Didden, W. A. M., 1990b. The Population Ecology and Functioning of Enchytraeidae in Some Arable Farming Systems, Doctoral thesis, Agricultural University, Wageningen, The Netherlands, p. 117.

Didden, W. A. M., Marinissen, J. C. Y., Vreeken-Bruijs, M. J., Burgers, S. L. G. E., de Fluiter, R., Geurs, M., and Brussaard, L., 1994. Soil meso- and macrofauna in two agricultural systems: factors affecting population dynamics and evaluation of their role in carbon and nitrogen dynamics. *Agric. Ecosys. Environ.,* 51:171–186.

Doran, J. W., 1980. Soil microbial and biochemical changes associated with reduced tillage. *Soil Sci. Soc. Am. J.,* 44:765–771.

Edwards, A. D. and Bremner, J. M., 1967. Microaggregates in soil. *J. Soil Sci.,* 18:64–73.

Elliott, E. T., 1986. Aggregate structure and carbon, nitrogen and phosphorus in native and cultivated soils. *Soil Sci. Soc. Am. J.,* 50:627–633.

Elliott, E. T., Coleman, D. C., and Cole, C. V., 1979. The influence of amoebae on the uptake of nitrogen by plants in gnotobiotic soil microcosms, in *The Soil-Root Interface,* Harley, J. L. and Russell, R. S., Eds., Academic Press, London, pp. 221–229.

Elliott, E. T., Anderson, R. V., Coleman, D. C., and Cole, C. V., 1980. Habitable pore space and microbial trophic interactions. *Oikos,* 35:327–335.

Elliott, E. T. and Coleman, D. C., 1988. Let the soil work for us. *Ecol. Bull.,* 39:23–32.

Elliott, E. T., Horton, K., Moore, J. C., Coleman, D. C., and Cole, C. V., 1984. Mineralization dynamics in fallow dryland wheat plots, Colorado. *Plant Soil,* 76:149–155.

Elliot, P. W., Knight, D., and Anderson, J. M., 1990. Denitrification in earthworm casts and soil from pastures under different fertilizer and drainage regimes. *Soil Biol. Biochem.,* 22:601–605.

Foster, R. C., 1981. Polysaccharides in soil fabrics. *Science,* 214:655–667.

Foster, R. C. and Dormaar, J. F., 1991. Bacterial-grazing amoebae in situ in the rhizosphere. *Biol. Fertil. Soils,* 11:83–87.

Freckman, D. W. and Mankau, R., 1986. Abundance, distribution, biomass and energetics of soil nematodes in a northern Mojave Desert ecosystem. *Pedobiologia,* 29:129–142.

Goering, H. K. and Van Soest, P. J., 1970. Forage Fiber Analyses, United States Department of Agriculture, Agriculture Handbook No. 379.

Golebiowska, J. and Ryszkowski, L., 1977. Energy and carbon fluxes in soil compartments of agroecosystems. *Ecol. Bull.* (Stockholm), 25:274–283.

Groffman, P. M., 1985. Nitrification and denitrification in conventional and no-tillage soils. *Soil Sci. Soc. Am. J.,* 49:329–334.

Gupta, V. V. S. R. and Germida, J. J., 1988. Distribution of microbial biomass and its activity in different soil aggregate size classes as affected by cultivation. *Soil Biol. Biochem.,* 20(6):777–786.

Hanlon, R. D. G. and Anderson, J. M., 1979. The effects of collembola grazing on microbial activity in decomposing leaf litter. *Oecologia,* 38:93–99.

Harper, S. H. T. and Lynch, J. M., 1985. Colonization and decomposition of straw by fungi. *Trans. Br. Mycol. Soc.,* 85:655–661.

Havlin, J. L., Kissel, D. E., Maddux, L. D., Claasen, M. M., and Long, J. H., 1990. Crop rotation and tillage effects on soil organic carbon and nitrogen. *Soil Sci. Soc. Am. J.,* 54:448–452.

Hendrix, P. F., Beare, M. H., Cheng, W., Parmelee, R. W., Coleman, D. C., and Crossley, D. A., Jr., 1989. Microbes, microfauna and movements of N-15 among litter, soil and plant pools in conventional and no-tillage agroecosystems. *Bull. Ecol. Soc. Am.,* 70(2):139.

Hendrix, P. F., Crossley, D. A., Jr., Coleman, D. C., Parmelee, R. W., and Beare, M. H., 1987. Carbon dynamics in soil microbes and fauna in conventional and no-tillage agroecosystems. *INTECOL Bull.,* 15:59–63.

Hendrix, P. F., Parmelee, R. W., Crossley, D. A., Coleman, D. C., Odum, E. P., and Groffman, P. M., 1986. Detritus food webs in conventional and no-tillage agroecosystems. *BioScience,* 36:374–380.

Holland, E. A. and Coleman, D. C., 1987. Litter placement effects on microbial and organic matter dynamics in an agroecosystem. *Ecology,* 68:425–433.

House, G. J. and Parmelee, R. W., 1985. Comparison of soil arthropods and earthworms from conventional and no-tillage agroecosystems. *Soil Tillage Res.,* 5:351–360.

Hunt, H. W., Coleman, D. C., Ingham, E. R., Ingham, R. E., Elliott, E. T., Moore, J. C., Rose, S. L., Reid, C. P. P., and Morley, C. R., 1987. The detrital food web in a shortgrass prairie. *Biol. Fertil. Soils,* 3:57–68.

Jackson, R. B. and Caldwell, M. M., 1993. The scale of nutrient heterogeneity around individual plants and its quantification with geostatistics. *Ecology,* 74:612–614.

Jastrow, J. D., 1987. Changes in soil aggregation associated with tallgrass prairie restoration. *Am. J. Bot.,* 74:1656–1664.

Jones, P. C. T. and Mollison, J. E., 1948. A technique for the quantitative estimation of soil microorganisms. *J. Gen. Microbiol.,* 2:54–69.

Killham, K., 1987. Heterotrophic nitrification, in *Nitrification,* Prosser, J., Ed., Society of General Microbiology, Spec. Public. I. R. L. Press, Oxford, pp. 117–126.

Killham, K., Sinclair, A. H., and Allison, M. F., 1988. Effect of straw addition on composition and activity of soil microbial biomass. *Proc. Royal Soc. Edinburgh,* 94B:135–143.

Kooistra, M. J., Lebbink, G., and Brussaard, L., 1989. The Dutch programme on soil ecology of arable farming systems. II. Geogenesis, agricultural history, field site characteristics and present farming systems at the Lovinkhoeve experimental farm. *Agric. Ecosystems Environ.,* 27:361–387.

Kuikman, P. J. and van Veen, J. A., 1989. The impact of protozoa on the availability of bacterial nitrogen to plants. *Biol. Fertil. Soils,* 8:13–18.

Lagerlöf, J., Andrén, O., and Paustian, K., 1989. Dynamics and contribution to carbon flows of Enchytraeidae (Oligochaeta) under four cropping systems. *J. Appl. Ecol.,* 26:183–199.

Laybourn-Parry, J., 1984. *A Functional Biology of the Free-Living Protozoa,* University of California Press, Berkeley, CA.

Lee, K. E., 1985. *Earthworms: Their Ecology and Relationships with Soils and Land Use,* Academic Press, New York, p. 411.

Lynch, J. M. and Harper, S. H. T., 1985. The microbial upgrading of straw for agricultural use. *Phil. Trans. R. Soc. London,* 310B:221–226.

Mallow, D. and Crossley, D. A., Jr., 1984. Evaluation of five techniques for recovering postlarval stages of chiggers from soil habitats. *J. Econ. Entomol.,* 77:281–284.

Marinissen, J. C. Y. and Dexter, A. R., 1990. Mechanisms of stabilization of earthworm casts and artificial casts. *Biol. Fertil. Soils,* 9:163–167.

Miller, R. M. and Jastrow, J. D., 1990. Hierarchy of root and mycorrhizal fungal interactions with soil aggregation. *Soil Biol. Biochem.,* 22:579–584.

Moore, J. C. and de Ruiter, P. C., 1991. Temporal and spatial heterogeneity of trophic interactions with below-ground food webs. *Agric. Ecosystems Environ.,* 34:371–397.

Moore, J. C., Zwetsloot, H. J. C., and de Ruiter, P. C., 1990. Statistical analysis and simulation modelling of the below-ground food webs of two winter-wheat management practices at the Lovinkhoeve (NL). *Neth. J. Agric. Sci.,* 38:303–316.

Mueller, B. R., Beare, M. H., and Crossley, D. A., Jr., 1990. Soil mites in detrital food webs on conventional and no-tillage agroecosystems. *Pedobiologia,* 34:389–401.

Neely, C. L., Beare, M. H., Hargrove, W. L., and Coleman, D. C., 1991. Relationships between fungal and bacterial substrate-induced respiration, biomass and plant residue decomposition. *Soil Biol. Biochem.,* 23:947–954.

O'Connor, F. B., 1955. Extraction of enchytraeid worms from a coniferous forest soil. *Nature,* 175:815–816.

Parkin, T. B., 1987. Soil microsites as a source of dentrification variability. *Soil Sci. Soc. Am. J.,* 51:1194–1199.

Parmelee, R. W. and Alston, D. G., 1986. Nematode trophic structure in conventional and no-tillage agroecosystems. *J. Nematol.,* 18:403–407.

Parmelee, R. W. and Crossley, D. A., Jr., 1988. Earthworm production and role in the nitrogen cycle of a no-tillage agroecosystem on the Georgia Piedmont. *Pedobiologia,* 32:353–361.

Parmelee, R. W., Beare, M. H., and Blair, J. M., 1989. Decomposition and nitrogen dynamics of surface weed residues in no-tillage agroecosystems under drought conditions: influence of resource quality on the decomposer community. *Soil Biol. Biochem.,* 21:97–103.

Parmelee, R. W., Beare, M. H., Cheng, W., Hendrix, P. F., Rider, S. J., Crossley, D. A., Jr., and Coleman, D. C., 1990. Earthworms and enchytraeids in conventional and no-tillage agroecosystems: A biocide approach to assess their role in organic matter breakdown. *Biol. Fertil. Soils,* 10:1–10.

Paustian, K., 1985. Influence of fungal growth pattern on decomposition and nitrogen mineralization in a model system, in *Ecological Interactions in Soil: Plants, Microbes and Animals,* Fitter, A. H., Atkinson, D., Read, D. J., and Usher, M. B., Eds., British Ecological Society Special Publication no. 4, Blackwell Scientific, Oxford, England, pp. 159–173.

Paustian, K., Andrén, O., Chlarholm, M., Hansson, A.-C., Johansson, G., Lagerlöf, J., Lindlerg, T., Pettersson, R., and Sohlenius, B., 1990. Carbon and nitrogen budgets of four agro-ecosystems with annual and perennial crops, with and without N fertilization. *J. Appl. Ecol.,* 27:60–84.

Persson, T., Bååth, E., Clarholm, M., Lundkist, H., Söderström, B. E., and Sohlenius, B., 1980. Trophic structure, biomass dynamics and carbon metabolism of soil organisms in a Scots pine forest, in *Structure and Function of Northern Coniferous Forests — An Ecosystem Study,* Persson, T., Ed., *Ecol. Bull.* (Stockholm), 32:419–459.

Peterson, H. and Luxton, M., 1982. A comparative analysis of soil fauna populations and their role in decomposition. *Oikos,* 39:287–388.

Ponge, J. F., 1991. Succession of fungi and fauna during the decomposition of needles in a small area of Scots pine litter. *Plant Soil,* 138:99–113.

Rasmussen, P. E. and Collins, H. P., 1991. Long-term impacts of tillage, fertilizer, and crop residue on soil organic matter in temperate semiarid regions. *Adv. Agron.,* 45:93–134.

Rice, C. W. and Smith, M. S., 1983. Nitrification of fertilizer and mineralized ammonium in no-till and plowed soils. *Soil Sci. Soc. Am. J.,* 47:1125–1129.

Robertson, G. P., 1994. The impact of soil and crop management practices on soil spatial heterogeneity, in *Soil Biota: Management in Sustainable Farming Systems,* Pankhurst, C. E., Doube, B. M., Gupta, V. V. S. R., and Grace, P. R., Eds., CSIRO, Australia.

Robertson, G. P., Huston, M. A., Evans, F. C., and Tiedje, J. M., 1988. Spatial variability in a successional plant community: patterns of nitrogen availability. *Ecology,* 69:1517–1524.

Robinson, C. H., Dighton, J., Frankland, J. C., and Coward, P. A., 1993. Nutrient and carbon dioxide release by interacting species of straw-decomposing fungi. *Plant Soil,* 151:139–142.

Rusek, J., 1985. Soil microstructures — contributions of specific organisms. *Quaest. Entomol.,* 21:497–514.

Rutherford, P. M. and Juma, N. G., 1989. Dynamics of microbial biomass and soil fauna in two contrasting soils cropped to barley (*Hordeum vulgare* L.). *Biol. Fertil. Soils,* 8:144–153.

Ryszkowski, L., 1985. Impoverishment of soil fauna due to agriculture. *INTECOL Bull.,* 20:203–217.

Santos, P. F. and Whitford, W. G., 1981. The effects of microarthropods on litter decomposition in a Chihuahuan desert ecosystem. *Ecology,* 62:654–663.

Seastedt, T. R., 1984. The role of microarthropods in decomposition and mineralization processes. *Annu. Rev. Entomol.,* 29:25–45.

Singh, B. N., 1946. A method of estimating the number of soil protozoa, especially amoebae, based on their differential feeding on bacteria. *Ann. Appl. Biol.,* 33:112–119.

Söderström, B. E., 1977. Vital staining of fungi in pure cultures and in soil with fluorescein diacetate. *Soil Biol. Biochem.,* 9:59–63.

Sohlenius, B., Boström, S., and Sandor, A., 1987. Long-term dynamics of nematode communities in arable soil under four cropping systems. *J. Appl. Ecol.,* 24:131–144.

Stinner, B. R., Crossley, D. A., Odum, E. P., and Todd, R. L., 1984. Nutrient budgets and internal cycling of N, P, K, Ca and Mg in conventional, no-tillage, and old-field ecosystems on the Georgia piedmont. *Ecology,* 65:354–369.

Stout, J. D., 1980. The role of protozoa in nutrient cycling and energy flow. *Adv. Microb. Ecol.,* 4:1–50.

Struwe, S. and Kjøller, A., 1985. Functional groups of bacteria on decomposing ash litter. *Pedobiologia,* 28:367–376.

Swift, M. J., Heal, O. W., and Anderson, J. M., 1979. *Decomposition in Terrestrial Ecosystems,* Studies in Ecology Vol. 5, Blackwell Scientific, Oxford, U.K., p. 372.

Tisdall, J. M. and Oades, J. M., 1982. Organic matter and water stable aggregates in soils. *J. Soil Sci.,* 33:141–161.

van Veen, J. A. and Paul, E. A., 1979. Conversion of biovolume measurements of soil organisms, grown under various moisture tensions, to biomass and their nutrient content. *Appl. Environ. Microbiol.,* 37:686–692.

van Vliet, P. C. J., Beare, M. H., and Coleman, D. C., 1995. Population dynamics and functional roles of Enchytraeidae (Oligochaeta) in hardwood forest and agricultural ecosystems, in *The Functional Significance and Regulation of Soil Biodiversity,* Collins, H., Robertson, G. P., and Klug, M. Y., Eds., Kluwer Academic Press, Dordrecht, pp. 237–245.

Walter, D. E., 1987. Trophic behavior of "mycophagous" microarthropods. *Ecology,* 68:226–229.

Wessen, B. and Berg, B., 1986. Long-term decomposition of barley straw: chemical changes an ingrowth of fungal mycelium. *Soil Biol. Biochem.,* 18:53–59.

Widden, P., Howson, G., and French, D. D., 1986. Use of cotton strips to relate fungal community structure to cellulose decomposition rates in the field. *Soil Biol. Biochem.,* 9:59–63.

Zwart, K. B., Burgers, S. L. G. E., Bloem, J., Bouwman, L. A., Brussaard, L., Lebbink, G., Didden, W. A. M., Marinissen, J. C. Y., Vreeken-Buijs, M. J., and de Ruiter, P. C., 1994. Population dynamics in the belowground food web in two different agricultural systems. *Agric. Ecosys. Environ.,* 51:187–198.

# CHAPTER 4

# Roots as Sinks and Sources of Nutrients and Carbon in Agricultural Systems

M. van Noordwijk and G. Brouwer

## INTRODUCTION

Roots are a special category of "soil biota," being part of organisms that live in two very different types of environment, aboveground and belowground. As they belong to the organisms that form the main target of all agricultural interventions and are directly involved in the efficiency with which nutrients and water are used for crop production, root knowledge should be "at the root" of any discussion on sustainable agriculture. Yet, little direct attention is given to roots, possibly because of a lack of (1) clear concepts on what to look for and (2) easy (be it quick and dirty) methods to observe the relevant properties in actual root systems. The need for maximizing resource use efficiency and minimizing environmental pollution has given a new drive for root ecological research. Within the soil, roots are the main sinks for nutrients during the life of the crop, but may also play a role in immobilizing nutrients during the initial stages of their decomposition; subsequently, they will form a source of nutrients for soil organisms and future plants. The efficiency of roots as sinks for nutrients, both during and after their life, partly determines the conflict between environmental and production aims of agriculture.

For a long time the prevailing concept was "the more extensive the root system, the higher crop production." We now know that this is not true across the full range of agricultural production conditions; maximum plant production can be obtained with relatively small root systems if the daily water and nutrient requirements are met by technical means as in intensive horticultural production systems (Van Noordwijk and De Willigen, 1987). Roots are, however, directly involved in the efficiency of plants to use available water and nutrient reserves in the soil, and therefore in reducing negative side effects of agricultural production by leaching and losses to the atmosphere. As a first estimate, we may still expect that "the more extensive the root system, the higher may be the nutrient and water use efficiency" (Van Noordwijk and De Willigen, 1991). The possibility of obtaining greater resource use efficiency can only be realized if total supply of nutrients and water is regulated according to the crop demands and the attainable resource use efficiency. On a field scale, both resource supply and possible crop production show spatial variability. Inadequate techniques for dealing with this variation may reduce the resource use efficiency well below what is possible in the normally small experimental units considered for research (Van Noordwijk and Wadman, 1992).

Figure 1 shows a scheme of N flows in an arable crop of the temperate zone on the Northern Hemisphere, starting the calendar in spring. If the long-term organic matter balance is secured, system outputs (harvested products, leaching, and gaseous losses) must be balanced by inputs. In the northern temperate zone the main crop growing season normally has a rainfall deficit, and leaching (beyond the crop root zone) is mostly confined to autumn and winter. Denitrification may occur in summer, but in early autumn conditions are especially conducive for this process as relatively high temperatures are

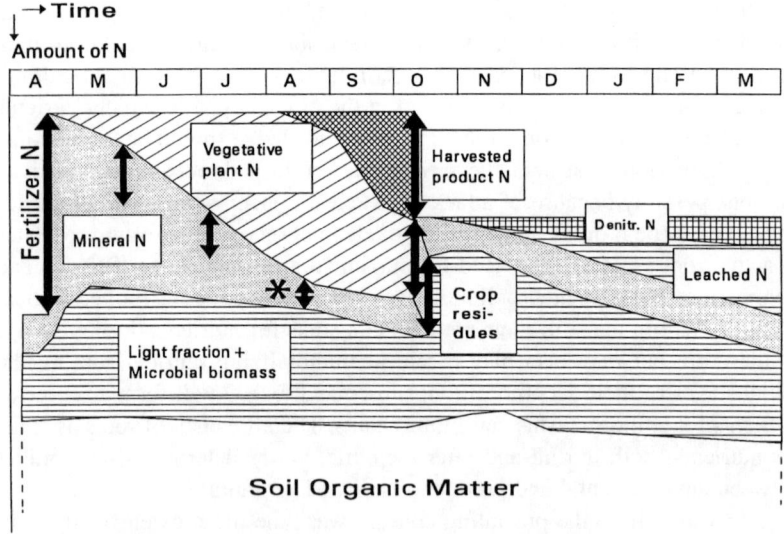

**Figure 1**   Scheme of N flows in an arable crop of the temperate zone on the Northern Hemisphere, starting in spring (April). The inputs as mineral fertilizer are partitioned over harvested products, crop residues, and soil mineral N at harvest time; losses to the environment by denitrification and leaching mainly occur during autumn and winter. The dynamics of the light fraction plus microbial biomass and its separation from the rest of soil organic matter are subject to debate and uncertainty. The arrows and asterisk indicate $N_{res}(N)$, the amount of mineral N needed to maintain sufficiently rapid diffusive transport toward the root system.

combined with wet soil conditions and ample availability of organic substrates. Rates of net N mineralization are less than required by crop demand during the growing season, but continue after the uptake period. Fertilizer requirements for avoidance of any N shortage in the crop have to allow for the weather-induced variability in crop demand and in rates of net N mineralization and also for the spatial variability within a field treated as a single management unit. Possibilities for "response farming" (i.e., tactical decision making during the growing season based on past and predicted weather conditions) are limited, however, as late fertilizer applications may not reach the roots under dry conditions. The arrows indicate that mineral N cannot be fully depleted, as a certain average concentration is needed to maintain diffusion gradients toward the root system, depending on soil water content, root length density, and current uptake demand (see below). The quantity corresponding with this concentration is termed $N_{res}$ for nutrients in general and $N_{res}(N)$ for mineral nitrogen. $N_{res}(N)$ is normally larger in the first phase of the growing season than later on, but most of this mineral N has to be present as it will be taken up anyway; $N_{res}(N)$ minus forthcoming N demand tends to be largest at the end of the uptake period (De Willigen and Van Noordwijk, 1987). The $N_{res}(N)$ at the end of the growing season (see asterisk) together with late mineralization, thus determines the minimum pool size from which N losses in winter occur. N fertilizer application from organic plus inorganic sources can be restricted to crop uptake plus this quantity $N_{res}(N)$ minus soil available N, but any further restriction will lead to yield loss. $N_{res}$ can be reduced by better root development. Manipulation of the light fraction plus microbial biomass pool to reduce net N mineralization after the uptake period can be an independent means of reducing N losses, for example, by adding residues with a high C:N ratio.

The relevance of various root parameters for predicting uptake efficiency depends not only on the resource studied but also on the complexity of the agricultural system. In intensive horticulture with nearly complete technical control over nutrient and water supply, fairly small root systems may allow very high crop productions in a situation where resource use efficiency ranges from very low to very high, depending on the technical perfection of the often soilless (hydroponic) production system (Van Noordwijk, 1990). In field crops grown as a monoculture, the technical possibilities are far lower for ensuring a supply of water and nutrients where and when needed by the crop, and it is difficult to achieve maximum crop production rates at high efficiency. The soil has to act as a buffer, temporarily storing these resources. Root systems are important in obtaining these resources as and when available and needed. Adjustment of supply and demand in both time and space (synchrony and synlocation) become critical factors. Critical values of root length density (root length per unit volume of soil) needed to obtain a specified water or nutrient use efficiency can be estimated from existing models. Normally, the rate-limiting step is formed by the transport from the soil matrix to the root surface and therefore by the geometry of the root-soil system and the required transport distances. In mixed cropping systems (including grasslands), the belowground interactions between the various plant species add a level of complexity to the system. On the one hand, mixed cropping systems allow complementarity in use of the space and thus of the stored resources, hence, improving overall resource efficiency. On the other hand, it means that root length densities that would be sufficient for efficient resource use in a monoculture may not be sufficient in a competitive situation. The plant mixtures found in agroforestry systems increase the complexity by another dimension, as the perennial and annual components have separate time frames on which to evaluate the interactions.

The soil water balance, as affected by climate, irrigation, and drainage, has a major influence on the required root functions in uptake. As the main crop growing season normally has a rainfall deficit in the northern temperate zone, dry soil conditions hamper diffusive transport and thus increase the root length density required for uptake. A lack of synchrony between N mineralization and N demand, which would lead to a build up of mineral N in the topsoil, is not a real problem under these conditions, as the main leaching risk occurs after the growing season. The main problem for effective use of organic N pools in the temperate zone is that mineralization is too slow in spring. In the humid tropics, by contrast, with a net rainfall surplus during most of the growing season, any accumulation of mineral N will be leached rapidly from the topsoil to deeper layers. Under such conditions synchrony of N mineralization and N demand is essential for obtaining high N use efficiencies and reducing leaching (Van Noordwijk and De Willigen, 1991; Myers et al., 1994).

Roots are a poorly quantified source of carbon in the soil ecosystem, and widely different estimates are available in the literature, at least partly due to uncertainties and biases caused by the methods used. Fine root residues of annual crops at harvest time tend to be low in nutrient content, as annual crops generally remobilize nutrients toward the end of their growth cycle, redistributing them to the seeds or vegetative storage tissue. For perennial crops with a less pronounced phenology, root decay may be a continuous process during the growing season. During their decomposition by microorganisms, roots may form a temporary sink of N. Little is known on the timing of net N mineralization from root residues and its role in the synchrony between net N mineralization and demand by the following crop.

The following root functions will be discussed here:

- Roots as sinks of nutrients (especially N and P) during the growing season, with emphasis on the $N_{res}$ term
- Roots as sources of carbon and hence as sinks and sources of nutrients (mainly N) after their death

Concepts and methods will be briefly reviewed and data will be given that were obtained as part of the Dutch Programme on Soil Ecology of Arable Farming Systems in which a conventional (CONV) and an "integrated" system (INT; reduced use of fertilizer, pesticides and soil tillage) were compared (Brussaard et al., 1988; Kooistra et al., 1989; Lebbink et al., 1994).

## MODELS AND METHODS FOR ROOTS AS NUTRIENTS SINKS

A wide array of mathematical models has been developed to integrate the soil chemical, soil physical, and plant physiological processes involved in uptake (Barber, 1984; De Willigen and Van Noordwijk, 1987; Nye and Tinker, 1977). The models can be used for extrapolating to other soil and climate (rainfall) conditions. Major uncertainties remain in the biological interactions in the rhizosphere. The efficiency of a root system in taking up nutrients (or water) from a layer of soil depends on the root length density (including mycorrhizal hyphae), root radius, soil water content, and effective diffusion coefficient for the nutrient (or water).

### Model for Simple Root-Soil Geometry

De Willigen and Van Noordwijk (1987, 1991) derived, with simplified assumptions on root-soil geometry, an equation for $N_{res}$ as a function of root length density $L_{rv}$. This relation can be used to predict uptake efficiency from a single homogeneous layer or can be incorporated into dynamic uptake models from layered soils.

$$N_{res} = \frac{A(K_a + \theta)D_m^2 G(\rho, \nu)}{4H(a_1\theta + a_0)\theta D_0}$$  (1)

with the dimensionless root parameter $\rho$ defined as:

$$\rho = 2(\pi L_{rv} D_m^2)^{-0.5}$$  (2)

and the dimensionless function G defined as:

$$G(\rho, 0) = \frac{\rho^2}{8}\left[-3 + \frac{1}{\rho^2} + \frac{4\ln\rho}{\rho^2 - 1}\right]$$  (3)

where  A  = daily nutrient demand (kg ha$^{-1}$ d$^{-1}$),
$K_a$  = apparent adsorption constant (ml cm$^{-3}$),
$\theta$  = soil water content (ml cm$^{-3}$),
$a_1$ and $a_0$ = parameters describing decrease of effective diffusion coefficient with decreasing $\theta$,
H  = depth of soil zone considered (cm),
$D_0$  = diffusion coefficient of nutrient in free water (cm$^2$ d$^{-1}$), and
$D_m$  = root diameter used for model (cm).
The function G depends on $\rho$ and $\nu$, a dimensionless group based on the transpiration rate (incorporating the effects of mass flow); the version given here is for $\nu$ equal to zero (compare De Willigen and Van Noordwijk, 1987).

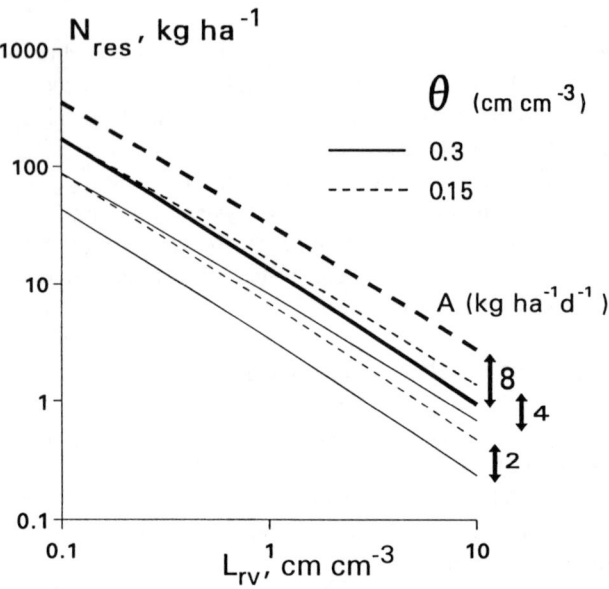

**Figure 2**  The amount of mineral, $N_{res}(N)$, in the soil (at two water contents $\theta$), required to maintain crop demand A, as function of root length density $L_{rv}$. (From De Willigen, P. and Van Noordwijk, M., 1987. Doctoral thesis, Agricultural University, Wageningen.)

Figure 2 shows $N_{res}(N)$ as function $L_{rv}$, A, and $\theta$ for a standard parameter set for $NO_3$ uptake (De Willigen and Van Noordwijk, 1987). $N_{res}(N)$ becomes less than 10 kg ha$^{-1}$ for $L_{rv}$ values in the range 0.4 to 3 cm cm$^{-3}$ (lower values for wetter soil and lower daily N demands). In view of the normal crop root length densities, we may thus expect that $N_{res}(N)$ in the topsoil can be small, except for high demands A on relatively dry soils (De Willigen and Van Noordwijk, 1987). Some of the simplifying assumptions, especially on the uniformity of root diameters and on the effects of root distribution pattern, can now be avoided by incorporating real-world complications into our scheme (Table 1).

## Heterogeneity in Root Diameter

If root systems of different diameters are compared at equal root length densities (length · diameter$^0$), the larger the diameter, the smaller is $N_{res}$ and thus the more efficient the uptake can be. If the comparison is made at equal surface area (length · diameter$^1$ · $\pi$), $N_{res}$ decreases with decreasing root diameter (De Willigen and Van Noordwijk, 1987). If the comparison is made at equal root volume (length · diameter$^2$ · $\pi/4$), or weight, the advantage of the smaller root diameters is even more pronounced. The most stable result (that means: a parameter least sensitive to changes in root diameter) is obtained

**Table 1    Steps in Describing Root-Soil Geometry for Uptake Models**

1.  Choose relevant sampling zones, based on depth, distance to crop rows, and expected synlocation of roots and resources; measure the root length density, $L_{rv}(i,s)$, for each stratum i at sample time s close to the expected maximum root development.

2.  Effective root length density for a root system with a known frequency distribution of root diameters (incl. hyphae):

$$L_{rv} = \frac{\sum_{j=1}^{n} L_{rv,j} \sqrt{D_j}}{\sqrt{D_m}} \qquad (5)$$

where $D_m$ = diameter used for model calculations and $D_j$ and $L_{rv,j}$ = root diameter and root length density of n diameter classes.

3.  Extrapolation from sampling time s to any time t is based on:

$$L_{rv}(i,t) = L_{rv}(i,s) \frac{R_p(i,t)}{R_p(i,s)} \qquad (6)$$

where $R_p(i,t)$ = relative root presence at zone i at time t on minirhizotron images.

$$R_p(i,t) = R_g(i,t) - R_d(i,t) \qquad (7)$$

where $R_g(i,t)$ = root growth in zone i till time t relative to year production and $R_d(i,t)$ = root decay in zone i till time t relative to year production.

4.  Measure effectiveness of the root distribution via the $R_{per}$ method (Van Noordwijk et al., 1993a,b) and derive the effective root length $L_{rv}$ · time t at zone i:

$$L_{rv}^{*}(i,t) = R_{per}(i,t) L_{rv}(i,t) \qquad (8)$$

where $R_{per}$ = root position effectivity ratio, accounting for nonregular root distribution and incomplete root-soil contact.

for a comparison at equal length · diameter$^{0.5}$. $N_{res}(P)$ can be translated into the required P availability in the soil as indicated by the (water-extractable) $P_w$ index. The more efficient the root system, the lower the required P level of the soil. Figure 3 shows that the required $P_w$ is least sensitive to variations in root diameter if root systems different in diameter are compared on the basis of equal root length · diameter$^{0.5}$. Calculations were made with the P model of Van Noordwijk et al. (1990) and were based on P adsorption isotherms and crop parameters for the growth of the velvet bean *Mucuna* on an Ultisol in Lampung, Indonesia (Hairiah et al., 1995).

With the root-root (length · diameter$^{0.5}$) index, calculation results are approximately independent of root diameter over more than one order of magnitude. We thus have a method to add hyphal length of mycorrhizal fungi (which are about a factor 25 smaller in diameter than the finest roots, based on a root diameter of 200 μm and a hyphal diameter of 8 μm) to the crop root length. Roughly one fifth (or $25^{0.5}$) of the hypha! length can be added to the

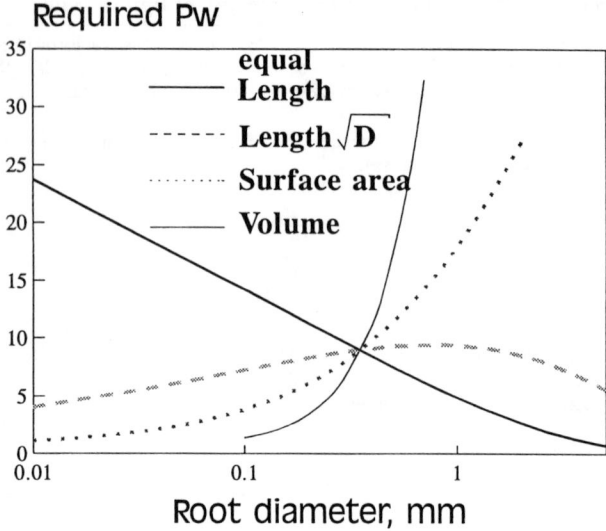

**Figure 3**  Required P availability in the soil — indicated by the (water extractable) $P_w$ index — when root systems of different diameter are compared on the basis of equal root length, root surface area, root volume, or sum of root length · diameter$^{0.5}$. The P model of Van Noordwijk et al. (1990) was applied to *Mucuna* growth on an Ultisol in Lampung (Hairiah et al., 1994) for these calculations.

root length density. If only "infection percentage" data are available for the mycorrhiza, we have to assume a reasonable length of hyphae per unit infected root length (a value between 10 and 100 seems reasonable, say 50) (Sanderson, personal communication, 1992). We thus obtain an increased root length density by a factor $1 + (0.5) \cdot \%inf/5$. For a normal infection percentage of 15%, this means that the effective root length density is 2.5 times the length of roots alone. There is an obvious lack of reliable data and relevant methods to quantify the hyphal length per unit infected root length, and this is clearly a priority area for research if process-based models for P uptake are desired. A similar but simpler method can be used to obtain a weighted average root diameter for a branched root system with diversity of root diameters (step 2 in Table 1).

## Nonregular Root Distribution

With the "root position effectivity ratio" $R_{per}$ the uptake efficiency for any actually observed root distribution pattern can be related to that for a theoretical, regular pattern. In an approximate manner, the effects of incomplete root-soil contact can also be incorporated (Van Noordwijk et al., 1993a,b). $R_{per}$ is defined as a reduction factor on the measured root length density and accounts for the lower uptake efficiency of real-world root distributions, when compared with the theoretical, regular pattern assumed by most existing uptake models (based on a cylinder geometry of the root-soil system), including the model used to derive Equation 1. For random root distributions, $R_{per}$ is approximately 0.5 (that means root length density X/2 in a regular pattern has the same $N_{res}$ as a random pattern at density X); the figures shown by Van Noordwijk et al. (1993a) are based on a different definition of $R_{per}$, later corrected by Van Noordwijk et al., (1993b). For clustered root distribution, as may be expected in structured soils, where roots grow mainly along cracks, $R_{per}$ values in the range 0.05 to 0.4 can be expected. $R_{per}$ tends to decrease with higher absolute root length densities.

## Dynamics of Root Growth and Decay

Due to the considerable spatial variability of root length density, estimates of $L_{rv}$ normally have a fairly wide confidence interval. If root growth and decay are estimated from a time series of destructive sampling, the results tend to have an unacceptably large uncertainty. If sequential nondestructive observations can be made on the same roots (e.g., those located next to a minirhizotron) and the resulting images are analyzed for changes relative to the root length present, a much smaller sampling error can be obtained. The cost of this, however, is a potential bias, as the observation method may influence root behavior, especially where gaps occur along the observation surface (Van Noordwijk et al., 1985). Details are given by Van Noordwijk et al. (1994a), who presented results of an analysis of sugar beet and winter wheat root turnover based on images obtained with inflatable minirhizotrons (Gijsman et al., 1991; Volkmar, 1993). Step 3 in Table 1 shows how the effective root length density at any time $t$ can be estimated from minirhizotron data plus a single destructive sampling.

## Effective Root Length Density as Function of Time and Depth

Combining these elements (Table 1), we can derive an effective root length density $L_{rv}*$ as a function of time and depth from

$$L_{rv}*(i, T) = R_{per}(i, T) \cdot \frac{\int_{t=0}^{T}(G_{i,t} - D_{i,t})dt}{\int_{t=0}^{s}(G_{i,t} - D_{i,t})dt} \cdot \frac{\sum_{j=0}^{n}L_{rv}(i,s,j)\sqrt{D_j}}{\sqrt{D_m}} \quad (4)$$

where $L_{rv}*(i,T)$ = effective root length density (cm cm$^{-3}$) in layer i at time T,

$\qquad L_{rv}(i,s)$   = measured root length density in layer i at time of sampling s,

$\qquad R_{per}(i,T)$ = root position effectivity ratio (procedure defined in Van Noordwijk et al., 1993b),

$\qquad G(i,t)$   = observed root growth along minirhizotrons as a function of time in zone i,

$D(i,t)$     = observed root decay along minirhizotrons as a function of time in zone i,

$D_m$     = root diameter used for model calculations, and

$D_j$     = root diameter for diameter class j and observed root length density $L_{rv}(j)$.

## METHODS FOR ROOTS AS SOURCE OF CARBON

Roots are a source of four forms of carbon: $CO_2$ from root respiration, soluble (exudates) C compounds, insoluble (mucigel, sloughed off root cap cells, etc.) C compounds released into the rhizosphere (during or after the life of the root), and structural root tissue after root death. Root-derived carbon forms the basis of a separate "energy channel" for the belowground food web (Moore et al., 1988; De Ruiter et al., 1994). It differs from other organic inputs, such as aboveground crop residues and organic manures, not only in quantity and chemical quality, but also in timing and spatial distribution. The position of roots in the soil, partly in the soil matrix and partly in larger aggregates or cracks, differs from the spatial distribution of other organic inputs, which occur in clusters and clumps, to a greater or lesser degree depending on the soil tillage operations used. Relatively stable organic matter-soil linkages are formed in the maize rhizosphere by mucigels from plant and bacterial origin (Watt et al., 1993). $^{14}C$ pulse labeling techniques have been used to measure the three types of belowground C output from plants (Swinnen, 1994). Integration over the whole growing season is needed to give a reliable estimate. Many previous studies concentrated on young plants only, and this overestimates exudation losses when extrapolated to the whole season.

Maximum standing stocks of fine root biomass are 0.5 to 2.5 Mg ha$^{-1}$ of dry weight for most annual crops and 1.5 to 5 Mg ha$^{-1}$ for pastures and grassland. The values for forests are normally in the same range for fine roots; the perennial woody main root system differs probably more between species and forest types than the fine roots. Data on standing root biomass in forests and under agricultural land use in the tropics are scarce. No belowground equivalent of the aboveground estimation procedure based on diameters at breast height (DBH) is available yet. However, application of fractal branching models holds a promise for new developments in this field (Van Noordwijk et al., 1994b). With methods for quantifying the maximum standing stocks reasonably well established, the annual turnover of the fine roots is the main source of uncertainty.

The C input from structural root tissue consists of root residues at harvest time and turnover of structural root material during the growing season. Structural root residues, especially if they have a low N concentration where plants experience some N stress at the final growth stages, may immobilize N after harvest time and thus reduce N losses by leaching. Estimates of annual turnover (relative to the maximum standing stock) vary from 1 to 10 yr$^{-1}$. Measurements of root turnover in the past decades have suffered from a lack of adequate methods. The widely used peak-and-trough method (Anderson and Ingram, 1993) is unsuitable in tropical pastures without pronounced seasonality. Despite impressive amounts of samples analyzed, no reliable estimates of turnover could be derived from this method for pastures in the Amazon (Castilla, personal communication). The best current method is based on sequential analysis of images taken in minirhizotrons as described above.

## Litter Pots for Studying Root Decay

Decay of root tissue is a gradual process that starts during the life of the root. Root decomposition after crop harvest can be studied by incorporating known amounts of roots (at a natural root length density), sieved or handpicked from soil, into sieved (root-free) soil, which is repacked to a natural soil bulk density and placed back into the field. Ceramic pots can be used as they allow water content in the pot to follow that of the surrounding soil. Pots are recovered at regular times. The remaining intact root mass is collected on a sieve and dried for weighing and chemical analysis (Van Noordwijk et al., 1994a). Caution is needed in interpreting the results of such studies, however, as the immediate root environment, including root-soil contact, has been modified. A comparison with decay along minirhizotron tubes (in a nonnatural but constant environment) is needed.

## RESULTS

## Root Length Density at Maximum Root Development

Figure 4 shows the distribution of root length density with depth at approximately maximum root development for winter wheat and sugar beet at the Lovinkhoeve experimental farm, The Netherlands. Root distribution differs distinctly from the normal exponential decrease of $L_{rv}$ with depth (which would lead to a straight line on the logarithmic scale used). A sandy layer at around 30 to 35 cm depth restricts root development, but below that zone root length density is fairly uniform up to about 80 cm depth. In sugar beet significant differences in $L_{rv}$ between three sampling positions (one in the crop row and two positions between the rows) persisted until a depth of 50 cm. Winter wheat root length densities were higher than those of sugar beet, throughout the profile. $L_{rv}$ in the INT treatment was slightly higher than that in the CONV treatment.

## Root Distribution Patterns

For sugar beet and winter wheat average values of $R_{per}$ obtained for the plough layer are 0.18 and 0.15, respectively (Figure 5). This suggests that a sixfold overestimate of the effective root length density is used if roots are

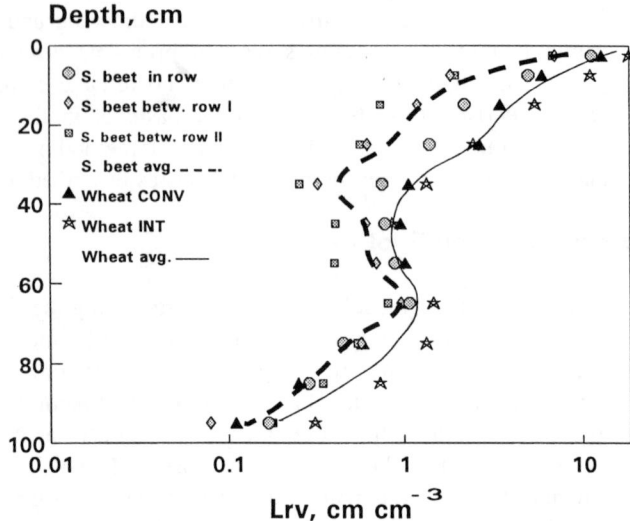

**Figure 4** Root length density $L_{rv}$ of winter wheat (late June 1990) and sugar beet (early September 1989) on the Lovinkhoeve experimental farm for the "conventional" (*CONV*) and "integrated" (*INT*) treatment. For sugar beet, two sample positions (I and II) were distinguished between the crop rows.

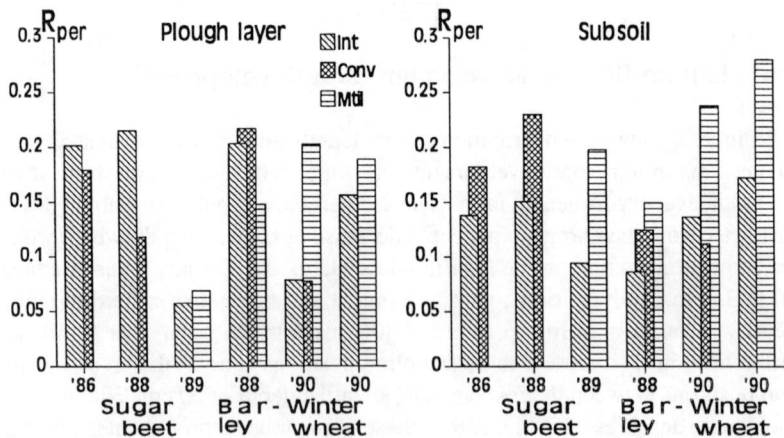

**Figure 5** Root position effectivity ratio, $R_{per}$, for sugar beet, spring barley, and winter wheat on the Lovinkhoeve experimental farm; observations were made on *CONV*, *INT*, and a reduced-tillage treatment (Mtil). Differences between crops and/or cropping systems were not statistically significant (the standard error of differences, S.E.D., for a comparison of crops is 0.03, for crop-system interactions 0.034).

assumed to be regularly distributed. Although no statistically significant differences between crops or cropping systems were found, the absolute value shows that root patterns deserve serious attention.

## Dynamics of Root Growth and Decay

For winter wheat and sugar beet grown in "conventional" and "integrated" arable cropping systems in The Netherlands, logistic-curve fits to growth and decay functions were given by Van Noordwijk et al. (1994a). The two crops showed a clearly different pattern, consistent over years of observation and largely independent of crop management regime. In winter wheat only little decay of roots was noticed during grain filling and ripening, and at harvest time 85 and 68% (in 1986 and 1990, respectively) of the structural root production remained as intact roots in the soil in both management systems. In sugar beet a more rapid and gradual root turnover was observed and the harvest residue of intact fine roots was on average 47% of cumulative root production. Figure 6 shows how absolute root length density can be estimated from the relative pattern observed on minirhizotrons and one destructive sampling (compare Table 1). The different depths and sample positions showed a fairly small variation in the growth curves, but a scatter in the root decay curves. Root decay was generally faster in the upper soil horizons (Van Noordwijk et al., 1994a).

## Total Structural Root Production

Cumulative root production of winter wheat was about 1700 kg ha$^{-1}$ for CONV and 1960 kg ha$^{-1}$ for INT crop management (Van Noordwijk et al., 1994a); in 1990 the difference between the two management systems was statistically significant. For sugar beet total fine root production was estimated at 1150 kg ha$^{-1}$ in 1987 and 1989, with a significantly lower amount on the field on which minimum tillage was introduced in 1986. Due to the higher turnover and smaller standing stock of sugar beet roots, the difference in total production of structural root material between winter wheat and sugar beet was much smaller than one would expect from single-point sampling.

## Roots Decay After Crop Harvest

Winter wheat root decay was studied with litter pots after crop harvest and in the following growing season (Figure 7). Initially, the N concentration in remaining roots increased while dry weight decreased. A slight net immobilization of minerals of N and P during autumn appeared in CONV but not in INT. During the next growing season the N concentration of remaining root debris was approximately constant and thus net mineralization was proportional to loss of root weight in an exponential decay with a half-life of 600°C days (daily temperature sum). This N release pattern during the next growing

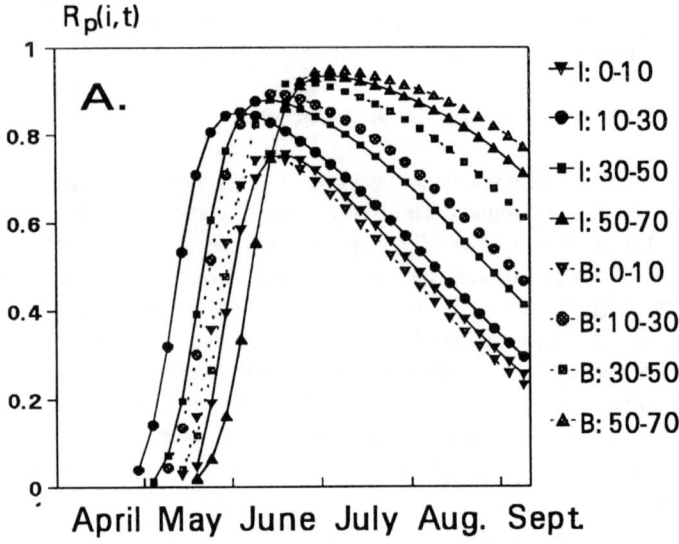

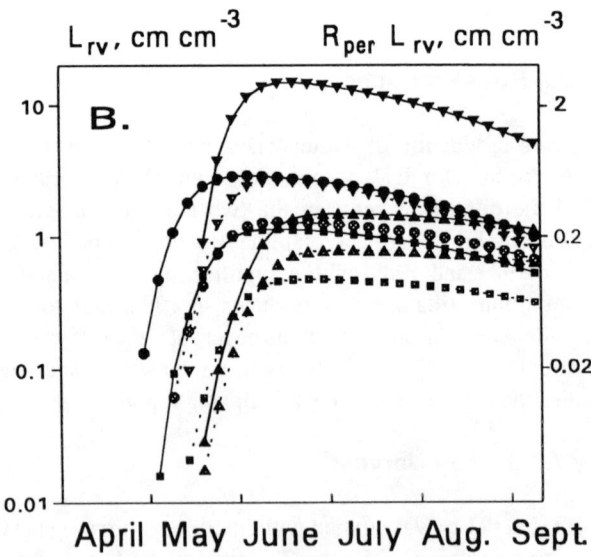

**Figure 6**  **A.** Relative root length as function of time, depth, and sample position (*I* = in row, *B* = between rows), as obtained from actual root length density observed in minirhizotrons divided by the cumulative year production, for sugar beet on INT in 1989. **B.** Estimated absolute root length density, on the basis of the relative pattern and one destructive sampling (log scale); the righthand Y scale estimates the effective root length density, based on an average root position effectivity ratio for sugar beet of 0.2.

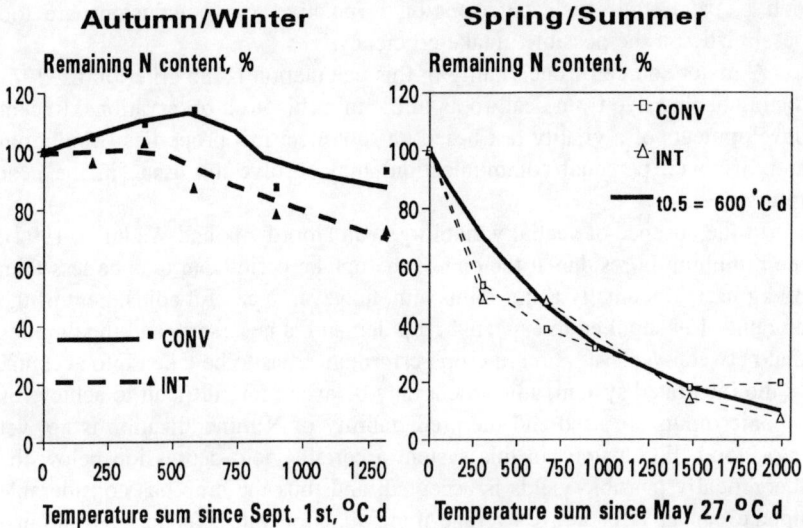

**Figure 7**    Remaining N content (concentration · dry weight, relative to the initial content) of decaying winter wheat roots in a litter pot decomposition study starting at harvest time (*left*) and one starting in May, with root residues collected in early spring.

period and the slight immobilization in CONV contribute to the synchrony between N demand and supply. Comparison of root decay along minirhizotrons and in the litter pots showed a strong effect of soil depth on root decay in the minirhizotrons (Van Noordwijk et al., 1994a).

## DISCUSSION

The combined data show that for both sugar beet and winter wheat the effective root length density in the topsoil, allowing for a $R_{per}$ of around 0.2, has been more than 0.2 cm cm$^{-3}$ throughout the later part of the growing season. As the soil at the Lovinkhoeve rarely dries out below a $\Theta$ of 0.2 ml cm$^{-3}$ (De Vos et al., 1994) and crop demand for N at the end of the uptake period will not have exceeded 2 kg ha$^{-1}$ d$^{-1}$, we may conclude that $N_{res}$ for mineral nitrogen probably has been in the range of 10 to 30 kg ha$^{-1}$ at the most critical point in time (compare Figure 1). In the wheat field both the higher root length density and the presence of mycorhiza will have led to $N_{res}(N)$ values of less than 10 kg ha$^{-1}$. We may thus tentatively conclude that root development of the nonmycorrhizal sugar beet is on the low side for efficient N uptake, at least if our correction for root distribution via $R_{per}$ is correct, while winter wheat has sufficient roots. The application of the $R_{per}$ correction assumes that there is no spatial correlation between roots and locally rich N patches; if such a correlation is positive, our conclusions are too pessimistic on the possible uptake efficiency.

A major source of uncertainty in this calculation is the criterion used for distinguishing live from dead roots in the minirhizotron observations. Recent developments of a vitality test based on root electrical properties (Meijboom and Brouwer, personal communication) may resolve this issue in the near future.

In the absence of spatial variability (Van Noordwijk and Wadman, 1992), the minimum N residue at the end of the uptake period can thus be less than 30 kg ha$^{-1}$. To actually achieve this aim, however, a careful adjustment of the amount of N supplied to expected crop demand is necessary, and this implies that between-year variation in crop performance has to be taken into account. In the integrated system, adjustment may be even more difficult to achieve as organic inputs are used and the predictability of N mineralization is not yet very good. Fortunately, in this system a certain yield depression below the theoretically possible yields is accepted, and thus the crop has considerable uptake capacity for above-average mineralization rates. In the conventional system, near-maximum yields are the goal and low N residues are difficult to obtain, despite the reliance on mineral N fertilizer instead of organic inputs.

An increase in soil mineral N content between the end of the uptake period and late autumn, when the N leaching season starts (De Vos et al., 1994), is to be expected due to late N mineralization. The initial N immobilization in decaying wheat root residues can have a small moderating effect on late N mineralization, but the total effect is probably less than 10 kg ha$^{-1}$ (Van Noordwijk et al., 1994a). Sugar beets are harvested so late that these effects will be small in that crop. Late N mineralization in sugar beet will greatly affect crop quality (higher amino acid content of the beets), but not necessarily increase N leaching, as the crop still has uptake capacity. The major limitations in achieving a high N uptake efficiency and low N losses by leaching thus are formed by post-harvest residue management, spatial variability in the field, and between-year variability in N demand, with root development in a secondary role, although the sugar beet data are in the critical range. For P the root length density of sugar beet observed here requires available P levels to be maintained, which is critical from an environmental point of view (Van Noordwijk et al., 1990).

Although it is a big step, we may try to extrapolate this conclusion to other conditions. Although the processes of the N balance and root growth and activity are the same in temperate zones and in the tropics, the rate at which they occur can be very different. The same crop root development and nutrient demand as in the Lovinkhoeve example would in the high-rainfall tropics lead to serious N loss during the growing season, unless the time pattern of N mineralization and N demand were closely similar and/or a considerable part of the mineral N stayed in the $NH_4^+$ form (higher $K_a$) instead of $NO_3^-$ (De Willigen and Van Noordwijk, 1987). On the acid soils, which are typical of the humid tropics, nitrification rates can be fairly slow, but root development is normally shallow, so N leaching losses during the growing season are still to be expected. Deep rooted (tree) components of mixed cropping systems can then act as a "safety-net," intercepting N on its way to deeper layers (Van Noordwijk and De Willigen, 1991). As a first step to a mechanistic approach to nutrient competition between different species in a mixture, we expect the competitive strength of each species to be (inversely) related to its $N_{res}$ value.

## ACKNOWLEDGMENTS

Thanks are due to two careful anonymous reviewers for helpful suggestions.

## REFERENCES

Anderson, J. M. and Ingram, J. S. I., 1993. *Tropical Soil Biology and Fertility: A Handbook of Methods,* 2nd ed. CAB International, Wallingford.

Barber, S. A., 1984. *Soil Nutrient Bioavailability: A Mechanistic Approach,* John Wiley & Sons, New York, p. 398.

Brussaard, L., Kooistra, M. J., Lebbink, G., and Van Veen, J. A., 1988. The Dutch programme on soil ecology of arable farming systems. I. Objectives, approach and some preliminary results. *Ecol. Bull.,* 39:35–40.

De Ruiter, P. C., Bloem, J., Bouwman, L. A., Didden, W. A. M., Hoenderboom, G. H. J., Lebbink, G., Marinissen, J. C. Y., De Vos, J. A., Vreeken-Buijs, M. J., Zwart, K. B., and Brussaard, L., 1994. Simulation of dynamics in nitrogen mineralization in the belowground food webs of two arable farming systems. *Agric. Ecosystems Environ.,* 51:199–208.

De Vos, J. A., Raats, P. A. C., and Vos, E. C., 1994. Macroscopic soil physical processes considered within an agronomical and a soil biological context. *Agric. Ecosystems Environ.,* 51:43–73.

De Willigen, P. and Van Noordwijk, M., 1987. Roots, Plant Production and Nutrient Use Efficiency, Doctoral thesis, Agricultural University, Wageningen, p. 282.

De Willigen, P. and Van Noordwijk, M., 1991. Modelling nutrient uptake: from single roots to complete root systems, in *Simulation and Systems Analysis for Rice Production* (SARP), Penning de Vries, F. W. T., van Laar, H. H., and Kropff, M. J., Eds., Simulation Monographs, PUDOC, Wageningen, pp. 277–295.

Gijsman, A. J., Floris, J., Van Noordwijk, M., and Brouwer, G., 1991. An inflatable minirhizotron system for root observations with improved soil/tube contact. *Plant Soil,* 134:261–269.

Hairiah, K., Van Noordwijk, M., and Setijono, S., 1995. Aluminum tolerance and avoidance of *Mucuna pruriens* at different P supply. *Plant Soil,* 171:77–81.

Kooistra, M. J., Lebbink, G., and Brussaard, L., 1989. The Dutch programme on soil ecology of arable systems. II. Geogenesis, agricultural history, field site characteristics and present farming systems at the Lovinkhoeve experimental farm. *Agric. Ecosystems Environ.,* 27:361–387.

Lebbink, G., Van Faassen, H. G., Van Ouwerkerk, C., and Brussaard, L., 1994. The Dutch programme on soil ecology of arable farming systems: farm management, monitoring programme and general results. *Agric. Ecosystems Environ.,* 51:7–20.

Moore, J. C., Walter, D. E., and Hunt, H. W., 1988. Arthropod regulation of micro- and mesobiota in below-ground detrital food webs. *Annu. Rev. Entomol.,* 33:419–439.

Myers, R. J. K., Palm, C. A., Cuevas, E., Gutatilleke, I. U. N., and Brossard, M., 1994. The synchronisation of nutrient mineralisation and plant nutrient demand, in *The Biological Management of Tropical Soil Fertility,* Woomer, P. L. and Swift, M. J., Eds., Wiley, New York, pp. 81–116.

Nye, P. H. and Tinker, P. B., 1977. *Solute Movement in the Soil-Root System.* Studies in Ecology No. 4, Blackwell, Oxford, p. 342.

Swinnen, J., 1994. Rhizodeposition and turnover of root-derived material in barley and wheat under conventional and integrated management. *Agric. Ecosystems Environ.,* 51:115–128.

Van Noordwijk, M., De Jager, A., and Floris, J., 1985. A new dimension to observations in mini-rhizotrons: a stereoscopic view on root photographs. *Plant Soil,* 86:447–453.

Van Noordwijk, M., 1990. Synchronization of supply and demand is necessary to increase efficiency of nutrient use in soilless horticulture, in *Plant Nutrition — Physiology and Applications,* van Beusichem, M. L., Ed., Kluwer Academic, Dordrecht, The Netherlands, pp. 525–531.

Van Noordwijk, M. and De Willigen, P., 1987. Agricultural concepts of roots: from morphogenetic to functional equilibrium. *Neth. J. Agric. Sci.,* 35:487–496.

Van Noordwijk, M. and De Willigen, P., 1991. Root functions in agricultural systems, in *Plant Roots and Their Environment,* Persson, H. and McMichael, B. L., Eds., Elsevier, Amsterdam, pp. 381–395.

Van Noordwijk, M. and Wadman, W., 1992. Effects of spatial variability of nitrogen supply on environmentally acceptable nitrogen fertilizer application rates to arable crops. *Neth. J. Agric. Sci.,* 40:51–72.

Van Noordwijk, M., De Willigen, P., Ehlert, P. A. I., and Chardon, W. J., 1990. A simple model of P uptake by crops as a possible basis for P fertilizer recommendations. *Neth. J. Agric. Sci.,* 38:317–332.

Van Noordwijk, M., Brouwer, G., and Harmanny, K., 1993a. Concepts and methods for studying interactions of roots and soil structure. *Geoderma,* 56:351–375.

Van Noordwijk, M., Brouwer, G., Zandt, P., Meijboom, F. J. M., and Burgers, S., 1993b. Root Patterns in Space and Time: Procedures and Programs for Quantification, IB-DLO Nota 268, Haren, The Netherlands, p. 122.

Van Noordwijk, M., Brouwer, G., Koning, H., Meijboom, F. J. M., and Grzebisz, W., 1994a. Production and decay of structural root material of winter wheat and sugar beet in conventional and integrated cropping systems. *Agric. Ecosystems Environ.,* 51:99–113.

Van Noordwijk, M., Spek, L. Y., and De Willigen, P., 1994b. Proximal root diameter as predictor of total root system size for fractal branching models. I. Theory. *Plant Soil,* 164:107–118.

Volkmar, K. M., 1993. A comparison of minirhizotron techniques for estimating root length density in soils of different bulk density. *Plant Soil,* 157:239–245.

Watt, M., McCully, M. E., and Jeffree, C. E., 1993. Plant and bacterial mucilages of the maize rhizosphere: comparison of their soil binding properties and histochemistry in a model system. *Plant Soil,* 151:151–165.

# CHAPTER 5

# Mycorrhizal Interactions with Plants and Soil Organisms in Sustainable Agroecosystems

J. Pérez-Moreno and R. Ferrera-Cerrato

## INTRODUCTION

Although farming has been affected by technological development, it remains basically an ecological enterprise. It is an activity in which natural ecosystems, open to the influence of climate, substrate, and wild biota, are modified to increase yields of desired food and fiber products. The greater the changes in the basic patterns of structure and function that prevail in the natural system, the greater is the human effort necessary to maintain the agricultural system (Cox and Atkins, 1979). Therefore, at the present time, it has been proved that conventional agriculture produces ecological disturbance and lack of sustainability, resulting in a reduction of soil fertility and increased damage by pathogens to cultivated plants. In addition, it has been observed that some traditional agricultural systems have higher sustainability and produce less ecological damage. These systems have been called low external-input agricultural (LEIA) systems. As opposed to the conventional systems, LEIA agroecosystems have high genetic and cultural diversity, multiple uses of resources, and efficient nutrient and material recycling (Altieri, 1987). The search for strategies to improve yields and to maintain these increases is a great challenge to human population at the present time (Pérez-Moreno and Ferrera-Cerrato, 1996). The role of biological alternatives, because of the intrinsic nature of farming, is of key importance to the search for reduced use of fertilizers, pesticides, and other chemicals. Among these alternatives, mycorrhiza management is particularly important because it strongly influences the plant nutrition processes and the soil stabilization.

1-56670-277-1/97/$0.00+$.50
© 1997 by CRC Press LLC

The mycorrhiza is a symbiotic association between some fungi and the roots of most plants (Brundett, 1991). Its physiological and ecological importance in natural ecosystems and its beneficial effects on cultivated plants have been widely documented (Marks and Kozlowski, 1973; Harley and Smith, 1983; Sieverding, 1991). It is well known that the characteristic dominant plants of each major terrestrial community associate mutualistically with soil fungi to form typical kinds of mycorrhiza. Therefore, ericaceous plants, which form the major component of the heathland biome, form with ascomycetous fungi distinctive "ericoid" mycorrhiza. In a similar way the dominant trees of the boreal and temperate forest biomes associate primarily with basidiomycetous fungi to form ectomycorrhiza, and the natural grasslands and most of the tropical rain forest species of the world form arbuscular mycorrhiza (AM) in association with fungi of zygomycetous affinities (Morton and Benny, 1990; Read, 1993). As natural ecosystem plants, most of the cultivated plants tend to form mycorrhizal associations. AM is the most widely distributed and colonizes most species of agricultural crops (Bethlenfalvay, 1992). The objective of this contribution is to discuss the importance of the mycorrhizal fungi and their interactions with plant management and other organisms in LEIA agroecosystems. In addition, the influence of some cultural practices on mycorrhizal fungi is discussed.

## STUDIES DEVELOPED IN LEIA SYSTEMS

### *Stizolobium*-Maize and Squash Rotation Agroecosystem

This is one of the main agroecosystems maintained for centuries in the tropical lowland, adjusted from the ecological and social viewpoints to maintain its productive capacity. It is based on culture rotation and polycultures. In addition, under this system no chemical fertilizer or pesticide is applied and no-tillage is carried out. It was described by Granados-Alvarez (1989) who pointed out that one of the main roles in the maintenance of the system is played by a plant locally called *nescafé* bean (*Stizolobium deeringianum* Bort.). This is a fast-growing legume that grows on the maize plants of the last harvest around April. In less than 2 months it covers the cultivated area entirely, eliminating the weed competition. The area is maintained in this condition for 7 to 8 months until November when, after its fructification, the *nescafé* is cleared with machetes. At this time the maize and squash are planted. The association grows well until the legume seeds begin to germinate; then they are cut with machetes. When the maize and squash harvest is complete, *Stizolobium* is left to grow freely. After harvest (March and April), the legume grows on the maize plants, covering again all the cultivated area and in this way closing the cycle. Some studies relating to the microbiology of the system have been carried out (González-Chávez et al., 1990a,b). A high (up to 80%) AM colonization has been reported. By contrast, maize monoculture colonization has been up to 50%. In addition, spore numbers, ranging from 8 to 400 spores $g^{-1}$ soil, have been observed. It is well known that AM fungi have their most significant effect on improving plant growth when little phosphate is present in the soil (Harley and Smith, 1983). If we take into account that the P concentration in the soils of tropical zones, like those of the *Stizolobium*-maize pumpkin agroecosystem, is very low (Galvis-Spinola, 1990), up to 4 to 7 μg $g^{-1}$ of P-Olsen (Quiroga-Madrigal, 1990), the soil around the maize-growing root is rapidly depleted of P ions within a distance of a few mm. Due to the extremely slow diffusion rate of P, this zone cannot be adequately replenished. However, direct uptake and transport of P by fungal hyphae have been confirmed by $^{32}P$ studies in other tropical agroecosystems (Sieverding, 1991). Thus external AM mycelium, which grows far beyond this depletion zone and increases the soil volume exploited for P uptake, may also contribute to the phosphorus nutrition of the plants grown in this agroecosystem.

In addition, it has been observed that plant clipping affects the AM colonization, reducing the abundance of arbuscles and increasing that of vesicles and spores (Vilariño and Arines, 1993). The length of AM external mycelium was also increased significantly with this treatment in comparison with control. In the described agricultural system, *Stizolobium* cutting could affect positively the AM production of storage of reserves and the production of resistant propagules and then produce the high observed AM incidence values. Different AM fungal species have been reported in the same tropical area in maize monoculture. Some species such as *Glomus constrictum* Trappe, *Acaulospora mellea* Spain et Schenck, and *Sclerocystis coccogena* (Pat.) Von Hohnel are present in the *Stizolobium*-maize rotation but not in the maize monoculture agroecosystem. This is important since differences between AM fungal species in altering host plant growth are well documented (Abbott and Robson, 1984; Bagyaraj, 1984; Chanway et al., 1991). The large number of parasite-infested and dead AM spores found reflects the intense symbiotic dynamics associated with the soil organisms in this agroecosystem. It has been reported that a wide variety of organisms, including nematodes and fungi (Siqueira et al., 1984; Williams, 1985; Ingham, 1988; Secilia and Bagyaraj, 1988), ingest, inhabit, or associate with hyphae or spores of AM fungi.

On the other hand, in this agroecosystem, multiple cropping presents higher $N_2$-fixing activity (Table 1). It has been observed that nitrogen fixation in legumes has an increase in activity during the vegetative period (Minchin et al., 1981). Subsequently, the time of flowering affects the amount of $N_2$ fixed, with the peak level of nitrogenase activity usually occurring during the early part of the reproductive stages when pods are still small (Bliss, 1987). In the discussed agroecosystem when *Stizolobium* was present, it followed this seasonal nitrogen-fixation profile. The highest nitrogenase activity was observed in the *Stizolobium* seed-filling stage, followed by a decline in subsequent periods (Table 1). It is well known that when seed filling begins in legumes, there is a great carbon sink affecting the supply of carbohydrates available for nodule growth, which is an important determinant of the total amount of $N_2$

**Table 1     Nitrogenase Activity in the *Stizolobium*-Maize and Squash Agroecosystem**

| Treatment | Season[a] | Ethylene produced (nmol g$^{-1}$ dry root h$^{-1}$) |
|---|---|---|
| *Stizolobium*-maize and squash agroecosystem with 9 years of management | I | 14 |
|  | II | 426 |
|  | III | 80 |
|  | IV | 116 |
| *Stizolobium*-maize and squash agroecosystem with 14 years of management | I | 76 |
|  | II | 243 |
|  | III | 86 |
|  | IV | 86 |
| Maize monoculture without *Stizolobium* or squash planting | I | 0 |
|  | II | 0 |
|  | III | 60 |
|  | IV | 0 |

[a] I, *Stizolobium* vegetative growth; II, *Stizolobium* sheat filling; III, maize flowering; IV, squash flowering.

Modified from González-Chávez et al., 1990b. *Agrociencia, Serie: Agua-Suelo-Clima.*, 1:133–153.

fixed. This explains the decline of nitrogenase activity during *Stizolobium* seed development observed in the agroecosystem.

## *Chinampas* Agroecosystem

Chinampas ("floating gardens") are agroecosystems that have maintained their ecological and productive sustainability for centuries. These systems have solved fertility and moisture problems using a simple technical manipulation. The agricultural system that produced the food for the Aztecs before the conquest by Spain in the Valley of Mexico is one of the most original and productive systems of agriculture known worldwide. At the present time, some areas surrounding Mexico City cultivate different agricultural products using this system. Polycultures are very commonly used, and there is year-round production of vegetables. Up to 28 vegetables along with maize are harvested each year in some chinampas. The agroecosystem has been described in detail by some authors (Coe, 1964; Armillas, 1971; Jiménez-Osornio and Núñez, 1993). Basically, it consists of farming plots constructed in swampy and shallow parts of a lake. The plot sides are reinforced with posts interwoven with branches and with willow trees planted along their edges. These plots are from 2.5 to 10 m wide and up to 100 m long creating a series of canals that separate the plots. Fertility is maintained by regular mucking and composting; at the present time plots are also manured. Special seedling nurseries using the sediments close to the plots are used. When appropriate, the bed of sediments is cut into blocks containing individual seedlings and these are transferred to the plots. In this way, fertility is always well balanced. Studies developed in Mexico of this system (Vera-Castello and Ferrera-Cerrato, 1990) showed that mycorrizal incidence appears to be low. In spite of this fact, the presence of AM fungi spores has been detected in the rhizospheric soil of some cultivated vegetables. The low incidence may be due to the rapid nutrient recycling through the addition and cycling of great amounts of green manure and soil sediments. In addition, the cultivation of nonmycorrhizal plants (such as Chenopodiaceae), which is a very common practice in the chinampas system, could affect mycorrhizal colonization. It has been observed that the roots of some of these species contain chemical factors inhibitory to mycorrhizal fungi (Tester et al., 1987). Another highly important factor that influences AM in this agroecosystem is water. It has been observed that flooded conditions affect negatively AM colonization and sporulation. In rice- and corn-based cropping systems the population of AM fungi is decreased after the wet season, when the field is inundated for a long period, and is increased in the dry season (Ilag et al., 1987). Solaiman and Hirata (1994) observed reductions from 6–33 to 0–4% in AM colonization and from 492–1,600 to 40–772 AM spores $kg^{-1}$, in wetland rice, caused by flooding. This could be caused by the influence of oxygen concentrations on AM. It has been shown that low oxygen concentrations (from 2 to 4%) in the soil atmosphere strongly reduce AM colonization (Saif, 1981, 1983).

However, it is well known (Lumsden et al., 1987, 1990; Zuckerman et al., 1989), that the conditions created under chinampas management produce suppression of damping-off caused by *Pythium* spp. and suppression of plant parasitic nematodes. Although there are few AM-colonized roots, the presence of other endorhizospheric fungi seems to be very frequent in the root system of the plants cultivated under this management according to our researches. These organisms play an important role in the biological control and plant growth. It has been observed that some of these fungi, such as *Trichoderma* spp., are capable of increasing plant growth and germination percentage rates and of creating short germination times for vegetables, and therefore they play a role as biocontrol agents (Harman et al., 1980; Kleifeld and Chet, 1992).

## *Marceño* Agroecosystem

Another important agroecosystem developed in tropical lowlands, including southern Mexico, is locally known as *marceño* (because it generally is planted in March). With this system 3 to 4 maize harvests per year are possible. This residual-moisture system has been practiced in areas flooded for 6 to 8 months per year where canals, raised platforms, and other structures that permit water manipulation have been constructed (Gliessman, 1991). In the dry season, as late as March, the wild vegetation or *popal* (mainly composed of aquatic plants as *Thalia geniculata* L.) is cleared and short-cycle varieties of maize are planted in the canals. When the maize plantlets have emerged, the system is set on fire. With this practice, weeds and other agents harmful for the culture are destroyed. The maize is harvested in June and July, before the flooding of the area. This cycle is repeated every year (Granados-Alvarez, 1989). In the raised platforms, planting is carried out in early June. At this time the conditions in the canals are too wet for planting. Harvest is carried out in late September. If the season is very wet, there is a second planting that is harvested in late January or early February (Gliessman et al., 1985).

Our researches have shown that AM colonization (up to 2%) and spore numbers in this agroecosystem are low, both in wild vegetation or maize stages. It has been observed (Dhillion et al., 1988; Vilariño and Arines, 1992; Dhillion and Anderson, 1993) that fire reduces AM propagule numbers and that the spores of some species from burned sites have lower germination rates than controls from neighboring unburned soil. In addition, it has been observed that the extracts of burned or heated soil reduce root colonization and arbuscle formation. It seems that burned soils contain water-soluble agents, reducing germination rates, AM colonization, arbuscle formation, and propagule density. This could explain the low AM incidence. However, our studies have shown that other microorganisms such as some $N_2$-fixing bacteria from the genera *Azospirillum, Derxia, Azotobacter,* and *Beijerinckia* are abundant in this agroecosystem (Table 2). With the *marceño* management some changes have been detected, reflecting the microorganism dynamics; for example, the number of actinomycetes has been significantly higher in cleared than in standing wild vegetation or the maize stage. The importance of these organisms in biological control is well known. If we take into account that these higher populations are present when maize is planted, we could consider its importance in pathogen control at the plantlet stage, which is when mainly root pathogens devastate maize in other tropical regions with conventional agriculture. In the meantime, $N_2$-fixing bacteria follow different dynamics, but all have high populations at the maize stage, playing an important role in the plant nutrition of the culture. In addition, as in *chinampas* soils, it seems that the presence of other endomycorrhizal fungi is common. These organisms, also are affecting pathogen damage, because it is well known that *marceño* soil also suppresses root pathogens such as *Pythium* (Lumsden et al., 1987, 1990; García-Espinosa, 1994).

Table 2   Abundance of Microorganisms in *Marceño* Agroecosystem

| Agroecosystem stage | (Colony-forming units g⁻¹ dry soil × 10³) | | | | | |
| --- | --- | --- | --- | --- | --- | --- |
| | TB | A | D | B | Az | F |
| Stood wild vegetation | 3600 a | 98 b | 6.8 a | 4.8 a | 3.6 a | 1.7 b |
| Cleared wild vegetation | 4200 a | 175 a | 5.0 b | 2.8 b | 3.2 a | 2.7 a |
| Maize cultivation (rhizosphere) | 3733 a | 71 b | 6.7 a | 3.0 b | 4.2 a | 2.8 a |

*Note:* TB, total bacteria; A, actinomycetes; D, *Derxia*; B, *Beijerinckia*; Az, *Azotobacter*; F, fungi. Values with the same letter in the same column are not different significantly (Tukey α = 0.05).

## Other LEIA Agroecosystems

Douds et al. (1992) studied the changes occurring in populations of AM fungi in two LEIA systems after 10 years of farming. These systems consisted of a LEIA maize-soybean rotation with animal manure as fertilizer and an emphasis on the production of hay, as well as grains, and a LEIA system with green manure and small-grain cover crops, which produce grain for income. When compared with a conventional maize-soybean rotation with chemical fertilizer and weed control, LEIA systems tended to have greater diversity and higher populations of spores of AM fungi than conventionally farmed plots. Some species such as *Gigaspora gigantea* (Nicolson et Gerdemann) Gerdemann et Trappe tended to be up to 30 times less common under conventional management than in LISA systems. *Glomus* spp. were also more numerous in the LISA systems. In addition, soil from these LISA systems produced greater colonization than from conventional systems in greenhouse bioassays with maize or Bahia grass (*Paspalum notatum* Flügge). As a result, the benefits of mycorrhizae were more conspicuous in these LISA systems.

In different areas of subtropical and temperate America some tree species are grown within agricultural crops such as maize. It has been observed that these trees influence soil fertility (Farrell, 1990). One of the most commonly used species is the capulin (*Prunus capuli* L.), which is endemic to Mexico. In agroecosystems where this tree species grows available phosphorus increases four- to sevenfold under the trees, and total carbon and potassium increase two- to threefold. Furthermore, nitrogen, calcium, and magnesium increase one-and-a-half to threefold, and cation exchange capacity increases one-and-a-half to twofold. Physical properties such as soil structure are also enhanced in these agroecosystems, developing more stable soil aggregates (Farrell, 1987). At the same time it has been observed that *P. capuli* is a highly mycorrhizal-dependent species. Inoculated plants have produced increments up to 1500% in dry weight with respect to uninoculated plants. Similar increments in almost any evaluated parameter, including plant height, stem diameter, leaf number, foliar area, and radical volume, have been found in plants inoculated with different AM fungi, including *Glomus aggregatum* Schenck et Smith emend. Koske, *G. fasciculatum* (Thaxter) Gerdemann et Trappe emend. Walker et Koske, *G. intraradix* Schenck et Smith, *Gigaspora margaria* Becker et Hall, and *Glomus* spp. (Jaen and Ferrera-Cerrato, 1989; Gómez and Ferrera-Cerrato, 1990; González-Cabrera et al., 1993). Taking into account their highly beneficial action, these results show that AM fungi also play an important role in the maintenance of these agroecosystems.

One of the typical features of a great number of LEIA agroecosystems is their great biological diversity, e.g., the "home garden" in tropical and subtropical regions of the world where crops, trees, and animals are combined in agroforestry systems, using the ecological structure of tropical rain forests to maintain a great diversity of products throughout the year. In these systems up to 80 plant species have been observed in 0.1 ha (Gliessman, 1990). If we take into account that there is a relation between plant and fungal diversity, systems like these have high AM fungal populations. It has also been observed that AM fungal diversity is negatively influenced in agricultural systems with high external inputs (fertilizers, pesticides, etc.) in tropical zones, while LEIA systems maintain medium to high diversity (Sieverding, 1990).

It is important to point out that in general terms the observed increases caused by AM fungi in the field have been smaller than in pot experiments, and some inconsistencies have been found. Fitter (1985) has considered that these may be due to (1) widespread distributions of AM ineffective strains (or species), (2) dissipation of benefits caused by interplant connections made by AM mycelium, (3) grazing of external hyphae by soil fauna, and (4) longevity of AM roots. Nevertheless, the agricultural use of AM may be possible if the effects of other organisms on mycorrhizal fungi could be modified to improve AM function, e.g., the grazing of soil fauna or the increase of populations of mycorrhization helper bacteria (Fitter and Garbaye, 1993). In addition, AM fungi are implicated in soil conservation via their role in soil aggregation (Miller and Jastrow, 1992). It has been shown (Tisdall, 1991) that networks of AM hyphae are important in binding microaggregates (0.02 to 0.25 mm diameter) into stable macroaggregates (>0.25 mm diameter). Electron microscopy studies (Gupta and Germida, 1988) have shown the importance of fungal hyphae for this macroaggregate formation. Because of their symbiotic nature and their persistence in the soil for several months after plants have died (Lee and Pankhurst, 1992), they have particular significance as stabilizers of soil aggregates. Indeed, it is believed that most of the microbial filaments that have been reported to stabilize aggregates in the field in the presence of plants are AM fungi (Tisdall and Oades, 1982). Also, mycorrhizal associations have been thought to play other important roles in the field: (1) in agrosystem regulation as a major interface or connection between the soil and plant subsystems (Bethlenfalvay, 1992) and (2) in improvement of both microbial and plant functions by acting mainly as transporters of mineral nutrients to the plant and C compounds to the soil biota (Bethlenfalvay and Linderman, 1992; Pérez-Moreno, 1995).

## CULTURAL PRACTICES COMMONLY USED IN LEIA SYSTEMS AND THEIR EFFECT ON MYCORRHIZAL FUNGI AND RELATED ORGANISMS

### No- or Reduced-Tillage

A key attribute of the AM is the production of a mycelial network, supported by the established plants, and hence a very high inoculum potential (Read, 1993). Hyphae play an important role in the formation, functioning, and perpetuation of mycorrhizas in agricultural ecosystems. Hyphae in soil, originating from either an established hyphal network or from other propagules (spores, vesicles, and root pieces), lead to the infection and subsequent colonization of roots (Abbott et al., 1992). In addition, there is evidence that AM hyphae can spread at least 11 cm from the roots (Li et al., 1991; Jakobsen et al., 1992a,b). However, the roles of the hyphae in phosphate uptake and soil stabilization are dependent on their distribution within the soil matrix in relation to the root surface. It has been observed that disturbance of the AM mycelial network negatively influences the plant growth and retards infection (Mulligan et al., 1985; Fairchild and Miller, 1988; Evans and Miller, 1988, 1990). In addition, it seems that the increased absorption of P caused by AM when soil is left undisturbed is due, at least in part, to the ability of the preexisting extraradical mycelium to act as a nutrient acquisition system for the newly developing plant. Indeed, the AM extraradical mycelium remains viable and retains its effectiveness as a nutrient acquisition system from one growing season to the next (Miller and McGonigle, 1992), and root fragments can also retain infectivity over periods of at least six months of storage (Tommerup and Abbott, 1981). At the same time, the hyphae of some AM species remain infective in soil dried to $-21.4$ MPa for at least 36 days (Jasper et al., 1989). The significance of this is that if the AM mycelium is left undisturbed under no- or reduced tillage, management will be able to facilitate both rapid infection and effective nutrient capture in environments with low fertility.

## Intercropping

Intercropping is the most common and most popular cropping system in Africa, Asia, and Latin America. On these continents 80% or more of the smallholder farmers grow two or more crops in association. The number of crops in the mixture can vary from two to a dozen, especially near the homestead (Edje, 1990). Although there are many complex combinations of intercrops, the predominant ones are simple and usually combine a cereal with a legume, grown as nutritional complements (Ofori and Stern, 1987). It has been estimated that high proportions of basic cereals are produced in multiple-crop systems in many parts of the world, including 90% of beans in Colombia, 80% of beans in Brazil, and 60% of maize in all the Latin American tropics. Whatever the crop combinations, intercropping is an intensive and sustainable land use system that the farmers have evolved over generations through experimentation (Francis, 1989).

Because many commonly occurring intercrop systems involve nodulating legumes, and since they frequently yield better than their monocultural components, it has been suggested that the legumes added nitrogen to the soil for the system as a whole, including transfer to the nonlegume plants (Vandermeer, 1989). It is conceivable that nitrogen is excreted by the legume roots into the soil (Brophy and Heichel, 1989; Wacquant et al., 1989) and is released as a normal decay process of nodules and roots (Haynes, 1980; Burity et al., 1989).

However, it has been proven that the more active mechanism involved is the AM transfer (Ames et al., 1983; Kessel et al., 1985; Francis et al., 1986; Haystead et al., 1988). Guzmán-Plazola et al. (1992) confirmed under field conditions that natural mycorrhizal links are established in intercrops between maize and bean. In addition, Kessel et al. (1985) confirmed the nitrogen transfer from soybean to maize plants. They used [15]N-labeled ammonium sulfate and 48 hours after application observed significantly higher values for atom percent [15]N excess in roots and leaves of AM-maize plants infected with *Glomus fasciculatum*. Also, it has been confirmed that compounds other than nitrogen may be transported from one plant to another through AM hyphal connections. There is strong evidence that [14]C can be transported between plants by mycorrhizal links (Brown et al., 1992). Other elements such as [45]Ca and [32]P are also believed to be transferred by this mechanism (Chiariello et al., 1982). However, there is no clear indication whether net transfer between linked plants ever occurs, and if so, whether the amount is large enough to benefit significantly the receiver plant. It is clear that when roots die, the transfer of phosphorus from one plant to another is increased by VA mycorrhizal links and that the amounts of nutrients involved are significant (Newman, 1988). Regarding this phenomenon, more recently Bethlenfalvay et al. (1991) pointed out that (1) AM-mediated N transfer from the root zone of soybean to maize varies with the mode of N input, (2) transfer of nutrients other than N is variable and can be significant and bidirectional, and (3) the direction of flow is related to source-sink relationships. Indeed, it seems that the effect of mycorrhizal fungi on soil microbial populations may be an important factor affecting N transfer between mycorrhizal plants, because high [15]N transfer from soybean to maize seems to be associated not only with high mycelium density but also with low soil microbial carbon (Hamel et al., 1991).

In addition, it has been observed that the exchange of root exudates between intercropped maize and bean without fertilization affects positively the effect of the mycorrhiza on plant growth (Guzmán-Plazola et al., 1992). These authors also observed that the endomycorrhizal fungi enhance the phosphorus and nitrogen absorption of maize and bean when they were intercropped. In spite of the higher levels of mycorrhizal colonization, maize showed lower effects to mycorrhizal inoculation than bean, providing evidence of the importance of nitrogen availability in the system functioning. In general terms, a bidirectional transfer in the AM fungus-host interfacial apoplast, very different from the mostly unidirectional flow in pathogens, has been suggested (Smith and Smith, 1986). Smith and Smith (1989) pointed out that the movement of P across active interfaces is thought to include active uptake of P by the fungus from the soil and loss from the fungus to the interface followed by active uptake by the root cells. This process would require changes in the efflux characteristics and loss of P from donor plant to the fungus at the interface. Although it cannot be assumed that the same mechanisms apply to all nutrients, it must be strongly emphasized that movement of a tracer from one place to another does not mean net transfer. Net fluxes depend on the relative fluxes in linked plants. Also, it has been reported that AM spores play an important role in introducing $N_2$-fixing bacteria such as *Acetobacter* into roots and shoots of cultivated plants (Boddey et al., 1991). In addition, AM fungi had a positive and highly significant effect on N fixation (Vejsadová et al., 1989; Azcón and Rubio, 1990; Reeves, 1992) contributing in this way also to enhanced plant nitrogen nutrition.

## Manure Addition and Other Practices

The addition of manure significantly stimulates AM frequency and intensity, but when applied together with N, P, and K, seems to cause a dramatic decrease in infection (Vejsadová, 1992). Studies developed in Mexico, in *tepetates* ("hardened soil layers") reclamation have shown that in polycultures the addition of bovine manure significantly increase AM colonization (Matías-Crisóstomo and Ferrera-Cerrato, 1993). However, the combined application of rock phosphate with animal manure increases the spore number per unit soil volume. These increments are up to 45%, and it seems that this effect is mainly due to improved reproduction of some species, including *G. fasciculatum, G. aggregatum,* and *G. geosporum* (Nicolson et Gerdemann) Walker (Heizemann et al., 1992). It has been demonstrated therefore that compounds present in animal dung and the slow release of P from rocks enforce the proliferation of Glomales in some tropical soils. In addition, some other practices commonly used in LISA agroecosystems as polyculture and terracing seem to favor AM. Some studies (Smith, 1980; Baltruschat and Dehne, 1988) have shown that a continuous monoculture adversely affects the inoculum potential of AM fungi. By contrast, it has been observed that AM infection and spore production increased in rotation with several cultures in relation to monoculture (Schenck and Kinloch, 1980; Sieverding and Leihner, 1984; Baltruschat and Dehne, 1989; Dodd et al., 1990). This could be related to the nutritious sources because polycultures seem to diversify their root exudates and then to promote higher biological diversity. It has also been observed that the culturing of highly mycorrhizal plants before other crops significantly increases AM colonization (Lippmann et al., 1990). However, cultivation of a nonhost crop in rotation with a host crop, or inclusion of a fallow period, may decrease spore numbers or propagule density of AM fungi in soils (Abbott and Robson, 1991). Mycorrhizal fungal communities are also affected by cropping history. Therefore, some species such as *Glomus aggregatum* Schenck are more abundant in soils with a corn history than a soybean history, while other species such as *G. albidum* Walker et Rhodes and *G. mosseae* Gerdemann et Trappe have the opposite trend (Johnson et al., 1991). With respect to terracing, it has been demonstrated that this practice in the tropical highlands of Africa enhances the presence of some AM fungi such as *Glomus callosum* Sieverding and *G. occultum* Walker (Heizemann et al., 1992). These variations of populations affect the AM root colonization and as a consequence the plant growth in these agroecosystems. As it has been discussed above, differential responses are produced according to the involved AM species.

## OTHER AGRICULTURAL PRACTICES CONVENTIONALLY USED AND THEIR EFFECT ON MYCORRHIZAL FUNGI

It has been shown that fertilizer application affects AM fungi. In spite of the complex interactions established among initial soil fertility, soil type (Hayman, 1982; Harley and Smith, 1983), organic matter content, and host plant and mycorrhizal fungi species, it seems that the factor that most strongly influences AM fungal colonization and sporulation is the P status of the plant (Kurle and Pfleger, 1994). It seems that in soils with very low P content, small amounts of phosphorus fertilizer do not affect AM colonization, whereas in soils with higher P levels, this kind of fertilization decreases infection (Johnson, 1984; Sieverding and Leihner, 1984; Douds et al., 1992; Vejsadová, 1992). Also, it has been observed that some plants only respond to AM inoculation in soils unamended with P fertilizer (Armstrong et al., 1992). In addition, a different ability to take up, translocate, and transfer phosphorus to the host plant according to the involved AM fungus has been reported (Pearson and Jakobsen, 1993). Other kinds of fertilization such as nitrogen do not seem to inhibit the symbiosis as do phosphorus or phosphorus plus nitrogen fertilization (Bentivenga and Hetrick, 1991). Indeed, increased AM spore numbers due to nitrogen addition have been reported (Bentivenga and Hetrick, 1992). However, an insufficient supply of nitrogen and its high doses cause a considerable decrease in colonization intensity (Gryndler et al., 1990). Also, it has been observed that Ca + Mg reduce the sporulation and increase the colonization (Anderson and Liberta, 1992). In general terms, high fertilizer application tends to reduce the AM fungi populations in tropical crops. Indeed, some AM species included in the *Sclerocystis* genus disappear when native systems are taken into agronomic plots (Sieverding, 1990). However, some species such as *Glomus manihot* Howeler, Sieverding et Schenck seem to tolerate different N, P, and K fertilizer application levels (Sieverding and Toro, 1990).

Pesticide application, a common and often obligatory practice in plant production, influences AM growth effects. These effects are differential according to the applied substances and can be beneficial or detrimental (Table 3). It has been shown that application rates and procedures also produce variable effects on the infection potential of inoculum and AM development (Parvathi et al., 1985). It has been established that pesticide effects, however, sometimes neutral or even positive (Plenchette and Perrin, 1992), usually decrease mycorrhizal infections and spore numbers (Ocampo and Hayman, 1980; Menge, 1982). Indeed, it has been shown that AM fungi may alleviate deleterious effects of some herbicides on plant growth when applied at low

Table 3     Influence of Some Pesticide Applications on Endomycorrhizal Plant
Growth Effects

| Pesticide | Effect | Source |
|---|---|---|
| Aldicarb | (+/−) | Nemec, 1985 |
| Aliette | (−) (+) (+/−) | Guillemin and Gianinnazzi, 1992; Morandi, 1990; Sukarno et al., 1993; Trouvelot et al., 1992 |
| Agrosan | (−) | Manjunath and Bagyaraj, 1984 |
| Benomyl | (−) | Parvathi et al., 1985; Trouvelot et al., 1992 |
| Benlate | (−) | Manjunath and Bagyaraj, 1984; Sukarno et al., 1993 |
| Captan | (+) (+/−) (−) | Guillemin and Gianinnazzi, 1992; Manjunath and Bagyaraj, 1984; Trouvelot et al., 1992 |
| Carbaryl | (−) (+/−) | Parvathi et al., 1985 |
| Ceresan | (−) | Manjunath and Bagyaraj, 1984 |
| Chlorpropham | (−) (+/−) | Ocampo and Barea, 1985 |
| Endosulfan | (−) (+/−) | Parvathi et al., 1985 |
| Etridiazole | (−) (+) | Guillemin and Gianinnazzi, 1992; Trouvelot et al., 1992 |
| Fenamiphos | (−) (+/−) | Nemec, 1985 |
| Fensulfothion | (−) | Nemec, 1985 |
| Furalaxyl | (−) | Trouvelot et al., 1992 |
| Imazaquin | (+/−) (−) | Siqueira et al., 1991 |
| Imazethapyr | (−) (+/−) | Siqueira et al., 1991 |
| Maneb | (+) | Guillemin and Gianinnazzi, 1992 |
| Parathion | (−) | Parvathi et al., 1985 |
| Pendimethalin | (−) | Siqueira et al., 1991 |
| Phenmedipham | (−) (+/−) | Ocampo and Barea, 1985 |
| Plantavax | (−) | Manjunath and Bagyaraj, 1984 |
| Plifenate | (+/−) | Nemec, 1985 |
| Propiconazole | (−) | Nemec, 1985 |
| Quintozene | (−) | Trouvelot et al., 1992 |
| Ridomil | (−) | Sukarno et al., 1993 |
| Sulfallate | (−) (+/−) | Ocampo and Barea, 1985 |
| Triadimefon | (+/−) | Nemec, 1985 |
| Trifoline | (+/−) | Nemec, 1985 |

Note: (−) detrimental, (+/−) neutral, (+) beneficial.

rates (Ocampo and Barea, 1985) because they stimulate isoflavonoid compound production (Siqueira et al., 1991).

## CONCLUSIONS

AM fungi play an important role in the nutrient uptake of many crops and are also associated with increased tolerance to water stress, decreased susceptibility to some plant diseases, and increased soil aggregation. Studies related with mycorrhizal research in LEIA agroecosystems have been scarce. However, it has been observed that mycorrhizal fungi play an important role in some of them. By contrast, in other sustainable agroecosystems, mycorrhizal incidence is low and other kinds of microorganisms are abundant and play important roles in the sustainability of the systems. So far, there has been insufficient investigation of interactions between AM and soil organisms in LEIA systems. Because of the high sustainability of these agroecosystems, it is suggested that much greater effort is required in the investigation of mycorrhizal symbiosis and their relationships with pathogenic fungi and soil fauna (mainly nematodes, collembola, and mites) in LEIA systems. In addition, the changes produced by AM in the rhizospheric functional groups such as actinomycetes and the influence of plant growth-promoting bacteria such as *Pseudomonas* on AM should also be studied for possible interactions, particularly synergistic effects. In addition, the selection of AM fungi with potential in soil aggregation must be deeply studied in the search of sustainability of agroecosystems. On the other hand, cultural practices frequently used in LEIA agroecosystems affect strongly rhizosphere microorganisms, including mycorrhizal fungi. In this way, practices such as no- or reduced tillage, intercropping, and crop rotation have been shown to favor the development of mycorrhiza. Other conventional agricultural practices such as pesticide or fertilizer application can stimulate or injure AM populations according to the dose and chemical nature of the substances. If we take into account that up to now the inoculation of these organisms has seemed to be more practical in plants that needed a nursery stage, better knowledge of the influence of cultural practices will optimize mycorrhizal management in the near future.

## ACKNOWLEDGMENT

We acknowledge Miss Sandra Aguilar-Sánchez for typing the manuscript and the valuable comments of an anonymous referee.

## REFERENCES

Abbott, L. K. and Robson, A. D., 1984. The effect of VA mycorrhizae on plant growth, in *VA Mycorrhiza,* Powell, C. L. and Bagyaraj, D. J., Eds., CRC Press, Boca Raton, FL, pp. 113–130.

Abbott, L. K. and Robson, A. D., 1991. Factors influencing the occurrence of vesicular-arbuscular mycorrhizas. *Agric. Ecosystems Environ.,* 35:121–150.

Abbott, L. K., Robson, A. D., Jasper, D. A., and Gazey, C., 1992. What is the role of VA mycorrhizal hyphae in soil?, in *Mycorrhizas in Ecosystems,* Read, D. J., Lewis, D. H., Fitter, A. H., and Alexander, I. J., Eds., C.A.B. International, Oxon, UK, pp. 37–41.

Altieri, M. A., 1987. Agroecology. The Scientific Basis of Alternative Agriculture. Westview Special Studies in Agriculture Science and Policy, Boulder, CO, p. 227.

Ames, R. N., Reid, C. P. P., Porter, L. K., and Cambardella, C., 1983. Hyphal uptake and transport of nitrogen from two $^{15}$N-labelled sources by *Glomus mosseae,* a vesicular arbuscular mycorrhizal fungus. *New Phytol.,* 95:381–396.

Anderson, R. C. and Liberta, A. E., 1992. Influence of supplemental inorganic nutrients on growth, survivorship, and mycorrhizal relationships of *Schizachyrium scoparium* (Poaceae) grown in fumigated and unfumigated soil. *Am. J. Bot.*, 79:406–414.

Armillas, P., 1971. Gardens on swamps. *Science*, 174:653–661.

Armstrong, R. D., Helyar, K. R., and Christie, E. K., 1992. Vesicular-arbuscular mycorrhiza in semi-arid pastures of south-west Queensland and their effect on growth responses to phosphorus fertilizers by grasses. *Aust. J. Agr. Res.*, 43:1143–1155.

Azcon, R. and Rubio, R., 1990. Interactions between different VA mycorrhizal fungi and *Rhizobium* strains on growth and nutrition of *Medicago sativa*. *Agric. Ecosystems Environ.*, 29:5–9.

Bagyaraj, D. J., 1984. Biological interactions with VA mycorrhizal fungi, in *VA Mycorrhiza*, Powell, C. L. and Bagyaraj, D. J., Eds., CRC Press, Boca Raton, FL, pp. 131–153.

Baltruschat, H. and Dehne, H. W., 1988. The occurrence of vesicular-arbuscular mycorrhiza in agro-ecosystems. I. Influence of nitrogen fertilization and green manure in continuous monoculture and in crop rotation on the inoculum potential of winter wheat. *Plant Soil*, 107:279–284.

Baltruschat, H. and Dehne, H. W., 1989. The occurrence of vesicular-arbuscular mycorrhiza in agroecosystems. II. Influence of nitrogen fertilization and green manure in continuous monoculture and in crop rotation on the inoculum potential of winter barley. *Plant Soil*, 113:251–256.

Bentivenga, S. P. and Hetrick, B. A. D., 1991. Relationship between mycorrhizal activity, burning, and plant productivity in tallgrass prairie. *Can. J. Bot.*, 69:2597–2602.

Bentivenga, S. P. and Hetrick, B. A. D., 1992. The effect of prairie management practices on mycorrhizal symbiosis. *Mycologia*, 84:522–527.

Bethlenfalvay, G. J., 1992. Mycorrhizae and crop productivity, in *Mycorrhizae in Sustainable Agriculture*, Bethlenfalvay, G. J. and Linderman, R. G., Eds., American Society of Agronomy, Crop Science Society of America and Soil Science Society of America, Madison, WI, pp. 1–27.

Bethlenfalvay, G. J. and Linderman, R. G., 1992. Preface, in *Mycorrhizae in Sustainable Agriculture*, Bethlenfalvay, G. J. and Linderman, R. G., Eds., American Society of Agronomy, Crop Science Society of America and Soil Science Society of America, Madison, WI, pp. viii–xiii.

Bethlenfalvay, G. J., Reyes-Solis, M. G., Camel, S. B., and Ferrera-Cerrato, R., 1991. Nutrient transfer between the root zones of soybean and maize plants connected by a common mycorrhizal mycelium. *Physiol. Plant.*, 82:423–432.

Bliss, F. A., 1987. Host plant control of symbiotic $N_2$ fixation in grain legumes, in *Genetic Aspects of Plant Mineral Nutrition*, Gabelman, H. W. and Loughman, B. C., Eds., Martinus Nijhoff Publishers, Dordrecht, pp. 479–493.

Boddey, R. M., Urquiaga, S., Reis, V., and Döbereiner, J., 1991. Biological nitrogen fixation associated with sugar cane. *Plant Soil*, 137:111–117.

Brophy, L. S. and Heichel, G. H., 1989. Nitrogen release from roots of alfalfa and soybean grown in sand culture. *Plant Soil*, 116:77–84.

Brown, M. S., Ferrera-Cerrato, R., and Bethlenfalvay, G. J., 1992. Mycorrhiza-mediated nutrient distribution between associated soybean and corn plants evaluated by the diagnosis and recommendation integrated system (DRIS). *Symbiosis*, 12:83–94.

Brundett, M., 1991. Mycorrhizas in natural ecosystems. *Adv. Ecol. Res.*, 21:171–313.

Burity, H. A., Ta, T. C., Faris, M. A., and Coulman, B. E., 1989. Estimation of nitrogen fixation and transfer from alfalfa to associated grasses in mixed swards under field conditions. *Plant Soil,* 114:249–255.

Chanway, C. P., Turkington, R., and Holl, F. B., 1991. Ecological implications of specificity between plants and rhizosphere micro-organisms. *Adv. Ecol. Res.,* 21:121–169.

Chiariello, N., Hickman, J. C., and Mooney, H. A., 1982. Endomycorrhizal role for interspecific transfer of phosphorus in a community of annual plants. *Science,* 217:941–943.

Coe, M. D., 1964. The chinampas of Mexico. *Sci. Am.,* 211:90–98.

Cox, G. and Atkins, M. D., 1979. *Agricultural Ecology. An Analysis of World Food Production Systems,* W.H. Freeman, San Francisco, p. 721.

Dhillion, S. S. and Anderson, R. C., 1993. Growth dynamics and associated mycorrhizal fungi of little bluesteam grass [*Schizachyrium scoparium* (Michx.) Nash] on burned and unburned sand prairies. *New Phytol.,* 123:77–91.

Dhillion, S. S., Anderson, R. C., and Liberta, A. E., 1988. Effect of fire on the mycorrhizal ecology of little bluesteam (*Schizachyrium scoparium*). *Can. J. Bot.,* 66:608–613.

Dodd, J. C., Arias, I., Koomen, K., and Hayman, D. S., 1990. The management of populations of vesicular-arbuscular mycorrhizal fungi in acid-infertile soils of a savanna ecosystem. *Plant Soil,* 122:229–240.

Douds, Jr., D. D., Janke, R. R., and Peters, S. E., 1992. Effect of 10 years of low-input sustainable agriculture upon VA fungi, in *Mycorrhizas in Ecosystems,* Read, D. J., Lewis, D. H., Fitter, A. H., and Alexander, I. J., Eds., C.A.B. International, Oxon, UK, p. 377.

Edje, O. T., 1990. Relevance of the workshop to farming in eastern and southern Africa, in *Research Methods for Cereal/Legume Intercropping,* Waddington, S. R., Palmer, A. F. E., and Edje, O. T., Eds., Proceedings of a Workshop on Research Methods for Cereal/Legume Intercropping in Eastern and Southern Africa, CIM-MyT, Mexico City, p. 5.

Evans, D. G. and Miller, M. H., 1988. Vesicular-arbuscular mycorrhiza and the soil disturbance-induced reduction of nutrient absorption in maize. I. Causal relations. *New Phytol.,* 110:67–74.

Evans, D. G. and Miller, M. H., 1990. The role of the external mycelium network in the effect of soil disturbance upon vesicular-arbuscular mycorrhizal colonization of maize. *New Phytol.,* 114:65–71.

Fairchild, G. F. and Miller, M. H., 1988. Vesicular-arbuscular mycorrhiza and the soil-disturbance-induced reduction of nutrient absorption in maize. II. Development of the effect. *New Phytol.,* 110:75–84.

Farrell, J. G., 1987. Agroforestry systems, in *Agroecology. The Scientific Basis of Alternative Agriculture,* Altieri, M. A., Ed., Westview Special Studies in Agriculture Science and Policy, Boulder, CO, pp. 149–158.

Farrell, J., 1990. The influence of trees in selected agroecosystems in Mexico, in *Agroecology: Researching the Ecological Basis for Sustainable Agriculture,* Gliessman, S. R., Ed., Springer-Verlag, New York, pp. 169–183.

Fitter, A. H. and Garbaye, J., 1993. Interactions between mycorrhizal fungi and other soil organisms. *Plant Soil,* 159:123–132.

Fitter, A. H., 1985. Functioning of vesicular-arbuscular mycorrhizas under field conditions. *New Phytol.*, 99:257–265.

Francis, R., Finlay, R. D., and Read, D. J., 1986. Vesicular-arbuscular mycorrhiza in natural vegetation systems. IV. Transfer of nutrients in inter- and intra-specific combinations of host plants. *New Phytol.*, 102:103–111.

Francis, C. A., 1989. Biological efficiencies in multiple-cropping systems. *Adv. Agron.*, 42:1–24.

Galvis-Spinola, A., 1990. Validación de las Normas de Fertilización de N y P Estimadas con un Modelo Simplificado Para Maíz, con las Dosis Obtenidas en la Experimentación de Campo, Master of Science thesis, Colegio de Postgraduados, Montecillo, Mexico, p. 113.

García-Espinosa, R., 1994. *Root Pathogens in the Agroecosystems of Mexico,* Transactions of the 15th World Congress of Soil Science, Volume 4a, International Society of Soil Science, Acapulco, Mexico, pp. 30–44.

Gliessman, S. R., 1990. Integrating trees into agriculture: the home garden agroecosystem as an example of agroforestry in the tropics, in *Agroecology: Researching the Ecological Basis for Sustainable Agriculture,* Gliessman, S. R., Ed., Springer-Verlag, New York, pp. 160–168.

Gliessman, S. R., 1991. Ecological basis of traditional management of wetlands in tropical Mexico: learning from agroecosystem models, in *Traditional Cultures and the Conservation of Biological Diversity,* Alcorn, J. and Oldfield, M., Eds., Westview Press, San Francisco, pp. 1–19.

Gliessman, S. R., Turner, B. L., Rosado-May, F. J., and Amador, M. F., 1985. Ancient raised field agriculture in the Maya lowlands of southern Mexico, in *Prehistoric Intensive Agriculture in the Tropics,* Farrington, I. S., Ed., BAR International Series 232, Oxford, pp. 97–111.

Gómez, C. G. and Ferrera-Cerrato, R., 1990. *Prunus serotina* sp. (*capuli* Cav. Mc Vaugh) y la Micorriza VA en Tepetate Amarillo del Estado de México, Proceedings Fifth Latinoamerican Congress of Botany, La Habana, p. 10.

González-Cabrera, V. H., González-Chávez, M. C., and Ferrera-Cerrato, R., 1993. Efecto de la mico riza vesículo-arbuscular en plantas de capulín (*Prunus serotina* var. capuli), in *Avances de Investigación. Sección de Microbiología de Suelos,* Pérez-Moreno, J. and Ferrera-Cerrato, R., Eds., Colegio de Postgraduados, Montecillo, State of Mexico, Mexico, pp. 92–99.

González-Chávez, M. C., Ferrera-Cerrato, R., and García-Espinosa, R., 1990a. *VA Mycorrhizae in a Sustainable Agro-Ecosystem in the Humid Tropics of Mexico,* 8th North Am. Conf. Myc., Jackson, WY, p. 120.

González-Chávez, M. C., Ferrera-Cerrato, R., García-Espinosa, R., and Martínez-Garza, A., 1990b. La fijación biológica de Nitrógeno en agroecosistemas de bajo ingreso externo de energía en Tamulté de las Sabanas, Tabasco. *Agrociencia, Serie: Agua-Suelo-Clima.*, 1:133–153.

Granados-Alvarez, N., 1989. La Rotación con Leguminosas Como Alternativa Para Reducir el Daño Causado por Fitopatógenos del Suelo y Elevar la Productivadad del Agroecoistema Maíz en el Trópico Húmedo, M.Sc., thesis, Colegio de Postgraduados, Montecillo, State of Mexico, Mexico, p. 111.

Gryndler, M., Leština, J., Moravec, V., Přikryl, Z., and Lipavsky, J., 1990. Colonization of maize roots by AM-fungi under conditions of long-term fertilization of varying intensity. *Agric. Ecosystems Environ.*, 29:183–186.

Guillemin, J. P. and Gianinnazzi, S., 1992. Fungicide interactions with VA fungi in *Ananas comosus* grown in a tropical environment, in *Mycorrhizas in Ecosystems,* Read, D. J., Lewis, D. H., Fitter, A. H., and Alexander, I. J., Eds., C.A.B. International, Wallingford, Oxon, UK, p. 381.

Gupta, V. V. S. R. and Germida, J. J., 1988. Distribution of microbial biomass and its activity in different soil aggregate size classes as affected by cultivation. *Soil Biol. Biochem.,* 20:777–786.

Guzmán-Plazola, R. A., Ferrera-Cerrato, R., and Bethlenfalvay, G. J., 1992. Papel de la endomicorrhiza V-A en la transferencia de exudados radicales entre frijol y maíz sembrados en asociación bajo condiciones de campo. *Terra,* 10:236–248.

Hamel, C., Nesser, C., Barrantes-Cartín, U., and Smith, D. L., 1991. Endomycorrhizal fungal species mediate $^{15}$N transfer from soybean to maize in non-fumigated soil. *Plant Soil,* 138:41–47.

Harley, J. L. and Smith, S. E., 1983. *Mycorrhizal Symbiosis,* Academic Press, London, p. 483.

Harman, G. E., Chet, I., and Baker, R., 1980. *Trichoderma hamatum* effects on seed and seedling disease induced in radish and pea by *Phythium* spp. or *Rhizoctonia solani. Phytopathology,* 70:1167–1172.

Hayman, D. S., 1982. Influence of soils and fertility on activity and survival of vesicular-arbuscular fungi mycorrhizal fungi. *Phytopathology,* 72:1119–1125.

Haynes, R. J., 1980. Competitive aspects of the grass-legume association. *Adv. Agron.,* 33:227–261.

Haystead, S., Malajczvk, N., and Grove, T. S., 1988. Underground transfer of nitrogen between pasture plants infected with vesicular-arbuscular mycorrhizal fungi. *New Phytol.,* 108:417–423.

Heizemann, J., Sieverding, E., and Diederichs, C., 1992. Native populations of the Glomales influenced by terracing and fertilization under cultivated potato in the tropical Highlands of Africa, in *Mycorrhizas in Ecosystems,* Read, D. J., Lewis, D. H., Fitter, A. H., and Alexander, I. J., Eds., C.A.B. International, Wallingford, Oxon, UK, p. 382.

Ilag, L. L., Rosales, A. M., Elazegui, F. A., and Mew, T. W., 1987. Changes in the population of infective endomycorrhizal fungi in a rice-based cropping system. *Plant Soil,* 103:67–73.

Ingham, R. E., 1988. Interactions between nematodes and vesicular-arbuscular mycorrhizae. *Agr. Ecos. Environ.,* 24:169–182.

Jaen, C. D. and Ferrera-Cerrato, R., 1989. Aplicación Tecnológica de los Hongos Endomicorrízicos en la Producción de Capulín (*Prunus serotina* var. *capuli*), Proceedings XXII Mexican Congress of Soil Science, Colegio de Postgraduados, Montecillo, State of Mexico, Mexico, p. 154.

Jakobsen, I., Abbott, L. K., and Robson, A. D., 1992a. External hyphae of vesicular-arbuscular mycorrhizal fungi associated with *Trifolium subterraneum* L. I. Spread of hyphae and phosphorus inflow into roots. *New Phytol.,* 120:371–380.

Jakobsen, I., Abbott, L. K., and Robson, A. D., 1992b. External hyphae of vesicular-arbuscular mycorrhizal fungi associated with *Trifolium subterraneum* L. II. Hyphal transport of $^{32}$P over defined distances. *New Phytol.,* 120:509–516.

Jasper, D. A., Abbott, L. K., and Robson, A. D., 1989. Hyphae of a vesicular-arbuscular mycorrhizal fungus maintain infectivity in dry soil except when soil is disturbed. *New Phytol.,* 112:101–107.

Jiménez-Osornio, J. J. and Núñez, P., 1993. La producción de chinampas diversificadas de San Andrés Mixquic, México, in *Agroecología, Sostenibilidad y Educación,* Ferrera-Cerrato, R. and Quintero-Lizaola, R., Eds., Colegio de Postgraduados, Montecillo, Estado de México, pp. 62–74.

Johnson, C. R., 1984. Phosphorus nutrition on mycorrhizal colonization, photosynthesis, growth and nutrient composition of *Citrus aurantium. Plant Soil,* 80:35–42.

Johnson, N. C., Pfleger, F. L., Crookston, R. K., Simmons, S. R., and Copeland, P. J., 1991. Vesicular-arbuscular mycorrhizas respond to corn and soybean cropping history. *New Phytol.,* 117:657–663.

Kessel, C. V., Singleton, P. W., and Hoben, H. J., 1985. Enhanced N-transfer from a soybean to maize by vesicular arbuscular mycorrhizal (AM) fungi. *Plant Physiol.,* 79:562–563.

Kleifeld, O. and Chet, I., 1992. *Trichoderma harzianum* — interaction with plants and effect on growth response. *Plant Soil,* 144:267–272.

Kurle, J. E. and Pfleger, F. L., 1994. The effects of cultural practices and pesticides on AM fungi, in *Mycorrhizae and Plant Health,* Pfleger, F. L. and Linderman, R. G., Eds., The American Phytopathological Society, St. Paul, MN, pp. 101–131.

Lee, K. E. and Pankhurst, C. E., 1992. Soil organisms and sustainable productivity. *Aust. J. Soil. Res.,* 30:855–892.

Li, X. L., George, E., and Marschner, H., 1991. Extension of the phosphorus depletion zone in VA-mycorrhizal white clover in a calcareous soil. *Plant Soil,* 136:41–48.

Lippmann, G., Witter, G., and Kegler, G., 1990. Factors controlling AM colonization percentage in arable soils. *Agric. Ecosystems Environ.,* 29:257–261.

Lumsden, R. D., García, R., Lewis, J. A., and Frías, G. A., 1987. Suppression of damping-off caused by *Pythium* spp. in soil from the indigenous Mexican chinampa agricultural system. *Soil. Biol. Biochem.,* 19:501–508.

Lumsden, R. D., García, R., Lewis, J. A., and Frias, T. G., 1990. Reduction of damping-off diseases in soil from indigenous Mexican agroecosystems, in *Agroecology, Researching the Ecological Basis for Sustainable Agriculture,* Gliessman, S. R., Ed., Springer-Verlag, New York, pp. 83–103.

Manjunath, A. and Bagyaraj, D. J., 1984. Effect of fungicides on mycorrhizal colonization and growth of onion. *Plant Soil,* 80:147–150.

Marks, G. C. and Kozlowski, T. T., Eds., 1973. *Ectomycorrhizae: Their Ecology and Physiology,* Academic Press, New York, p. 444.

Matías-Crisóstomo, S. and Ferrera-Cerrato, R., 1993. Efecto de microorganismos y adición de materia orgánica en la colonización micorrízica en la recuperación de tepetates, in *Avances de Investigación,* Pérez-Moreno, J. and Ferrera-Cerrato, R., Eds., Sección de Microbiología de Suelos, Colegio de Postgraduados, Montecillo, State of Mexico, Mexico, pp. 52–61.

Menge, J. A., 1982. Effect of soil fumigants and fungicides on vesicular-arbuscular fungi. *Phytopathology,* 72:1125–1132.

Miller, M. H. and McGonigle, T. P., 1992. Soil disturbance and the effectiveness of arbuscular mycorrhizas in an agricultural ecosystem, in *Mycorrhizas in Ecosystems,* Read, D. J., Lewis, D. H., Fitter, A. H., and Alexander, I. J., Eds., C.A.B. International, Wallingford, Oxon, UK, pp. 156–163.

Miller, R. M. and Jastrow, J. D., 1992. The role of mycorrhizal fungi in soil conservation, in *Mycorrhizae in Sustainable Agriculture,* Bethlenfalvay, G. J. and Linderman, R. G., Eds., American Society of Agronomy, Crop Science Society of America and Soil Science Society of America, Madison, WI, pp. 29–44.

Minchin, F. R., Summerfield, R. J., Hadley, P., Roberts, E. H., and Rawsthorne, S., 1981. Carbon and nitrogen nutrition of nodulated roots of grain legumes. *Plant Cell Environ.,* 4:5–26.

Morandi, D., 1990. Effect of endomycorrhizal infection and biocides on phytoalexin accumulation in soybean roots. *Agric. Ecosystems Environ.,* 29:303–305.

Morton, J. B. and Benny, G. L., 1990. Revised classification of arbuscular mycorrhizal fungi (Zygomycetes): a new order, Glomales, two new suborders, Glomineae and Gigasporineae, and two new families, Acaulosporaceae and Gigasporineae, and two new families, Acaulosporaceae and Gigasporaceae, with an emendation of Glomaceae. *Mycotaxon,* 37:471–491.

Mulligan, M. F., Smucker, A. J. M., and Safir, G. F., 1985. Tillage modifications in dry edible bean root colonization by AM fungi. *Agron. J.,* 77:140–144.

Nemec, S., 1985. Influence of selected pesticides on *Glomus* species and their infection in citrus. *Plant Soil,* 84:133–137.

Newman, E. I., 1988. Mycorrhizal links between plants: their functioning and ecological significance. *Adv. Ecol. Res.,* 18:243–270.

Ocampo, J. A. and Hayman, D. S., 1980. Effects of pesticides on mycorrhiza in field-grown barley, maize and potatoes. *Trans. Br. Mycol. Soc.,* 74:413–416.

Ocampo, J. A. and Barea, J. M., 1985. Effect of carbamate herbicides on VA mycorrhizal infection and plant growth. *Plant Soil,* 85:375–383.

Ofori, F. and Stern, W. R., 1987. Cereal-legume intercropping systems. *Adv. Agron.,* 41:41–90.

Parvathi, K., Ven Kateswarlu, K., and Rao, A. S., 1985. Effects of pesticides on development of *Glomus mosseae* in groundnut. *Trans. Br. Mycol. Soc.,* 84:29–33.

Pearson, J. N. and Jakobsen, I., 1993. The relative contribution of hyphae and roots to phosphorus uptake by arbuscular mycorrhizal plants, measured by dual labelling with $^{32}$P and $^{33}$P. *New Phytol.,* 124:489–494.

Pérez-Moreno, J., 1995. La simbiosis ectomicorrízia y su importancia ecológica, in *Agromicrobiología, Elemento útil en la Agricultura Sustentable,* Ferrera-Cerrato, R. and Pérez-Moreno, J., Eds., Colegio de Postgraduados, Montecillo, Mexico, pp. 200–233.

Pérez-Moreno, J. and Ferrera-Cerrato, R., Eds., 1996. *New Horizons in Agriculture: Agroecolocy and Sustainable Development,* Colegio de Postgraduados, Montecillo, Mexico, p. 435.

Plenchette, C. and Perrin, R., 1992. Evaluation in the green house of the effects of fungicides on the development of mycorrhiza on leek and wheat. *Mycorrhiza,* 1:59–62.

Quiroga-Madrigal, R. R., 1990. Impacto del Patosistema Edáfico del Maíz (*Zea mays*) en el Sistema de Rotación *Stizolobium* Maíz-Calabaza en Tamulté de las Sabanas, Tabasco, Master of Science thesis, Colegio de Postgraduados, Montecillo, Mexico, p. 125.

Read, D. J., 1993. Mycorrhiza in plant communities. *Adv. Plant Pathol.,* 9:1–29.

Reeves, M., 1992. The role of AM fungi in nitrogen dynamics in maize-bean intercrops. *Plant Soil,* 144:85–92.

Reid, C. P., 1990. Mycorrhizas, in *The Rhizosphere,* Lynch, J. M., Ed., Wiley, New York, pp. 281–315.

Saif, S. R., 1981. The influence of soil aeration on the efficiency of vesicular-arbuscular mycorrhizas. I. Effect of soil oxygen on the growth and mineral uptake of *Eupatorium adoratum* L. inoculated with *Glomus macrocarpus*. *New Phytol.*, 88:649–659.

Saif, S. R., 1983. The influence of soil aeration on the efficiency of vesicular-arbuscular mycorrhizas. II. Effect of soil oxygen on growth and mineral uptake in *Eupatorium odoratum* L., *Sorghum bicolor* (L.) Moench and *Guizotia abyssinica* (L.f.) Cass. inoculated with vesicular-arbuscular mycorrhizal fungi. *New Phytol.*, 95:405–417.

Schenck, N. C. and Kinloch, R. A., 1980. Incidence of mycorrhizal fungi on six field crops in monoculture on a newly cleared woodland site. *Mycologia*, 72:445–456.

Secilia, J. and Bagyaraj, D. J., 1988. Fungi associated with pot cultures of vesicular-arbuscular mycorrhizas. *Trans. Br. Myco. Soc.*, 91:117–119.

Sieverding, E., 1990. Ecology of AM fungi in tropical agroecosystems. *Agric. Ecosystems Environ.*, 29:369–390.

Sieverding, E., 1991. Vesicular arbuscular mycorrhizae management in tropical agroecosystems. Technical Cooperation, Federal Republic of Germany, Eschborn, p. 271.

Sieverding, E. and Leihner, D. E., 1984. Influence of crop rotation and intercropping of cassava with legumes on VA mycorrhizal symbiosis of cassava. *Plant Soil*, 80:143–146.

Sieverding, E. and Toro, S., 1990. Effect of mixing AM inoculum with fertilizers on Cassava nutrition and AM fungal association. *Agric. Ecosystems Environ.*, 29:397–401.

Siqueira, J. O., Hubbell, D. H., Kimbrough, J. W., and Schenk, N. C., 1984. *Stachybotrys chartarum* antagonistic to azygospores of *Gigaspora margarita* in vitro. *Soil Biol. Biochem.*, 16:679–681.

Siqueira, J. O., Safir, G. R., and Nair, M. G., 1991. VA-mycorrhiza stimulating isoflavonoid compounds reduce plant herbicide injury. *Plant Soil*, 134:233–242.

Smith, T. F., 1980. The effect of season and crop rotation on the abundance of spores of vesicular-arbuscular (VA) mycorrhizal endophytes. *Plant Soil*, 57:475–479.

Smith, F. A. and Smith, S. E., 1986. Movement across membranes: physiology and biochemistry, in *Physiological and Genetical Aspects of Mycorrhizae*, Gianinnazzi-Pearson, V. and Gianinnazzi, S., Eds., Institut National de la Recherche Agronomique, Paris, p. 75–84.

Smith, F. A. and Smith, S. E., 1989. Solute transport at the interface: ecological implications. *Agric. Ecosystems Environ.*, 28:475–478.

Solaiman, M. Z. and Hirata, H., 1994. *Ecophysiology of Vesicular-Arbuscular Mycorrhizal Fungi in Wetland Rice in Relation to Soil P and Water Regimes*, Transactions 15th World Congress of Soil Science, Volume 4b, Acapulco, Mexico, pp. 91–92.

Sukarno, N., Smith, S. E., and Scott, E. S., 1993. The effect of fungicides on vesicular-arbuscular mycorrhizal symbiosis. I. The effects on vesicular-arbuscular mycorrhizal fungi and plant growth. *New Phytol.*, 125:139–147.

Tester, M., Smith, S. E., and Smith, F. A., 1987. The phenomenon of "nonmycorrhizal" plants. *Can. J. Bot.*, 65:419–431.

Tisdall, J. M. and Oades, J. M., 1982. Organic matter and water-stable aggregates in soils. *J. Soil Sci.*, 33:141–163.

Tisdall, J. M., 1991. Fungal hyphae and structural stability of soil. *Aust. J. Soil Res.,* 29:729–743.

Tommerup, I. C. and Abbott, L. K., 1981. Prolonged survival and viability of VA mycorrhizal hyphae after root death. *Soil Biol. Biochem.,* 13:431–433.

Trouvelot, A., Abdel-Fattah, G. M., Gianinnazzi, S., and Gianinnazzi-Pearson, V., 1992. Differencial effects of fungicides on VA fungal viability and efficiency, in *Mycorrhizas in Ecosystems,* Read, D. J., Lewis, D. H., Fitter, A. H., and Alexander, I. J., Eds., C.A.B. International, Wallingford, Oxon, UK.

Vandermeer, J., 1989. *The Ecology of Intercropping,* Cambridge University Press, London, p. 357.

Vejsadová, H., 1992. The influence of organic and inorganic fertilization on development of indigenous VA fungi in roots of red clover, in *Mycorrhizas in Ecosystems,* Read, D. J., Lewis, D. H., Fitter, A. H., and Alexander, I. J., Eds., C.A.B. International, Wallingford, Oxon, UK, pp. 406–407.

Vejsadová, H., Hršelová, H., Přikryl, Z., and Vančura, V., 1989. Effect of different phosphorus and nitrogen levels on development of VA mycorrhiza, rhizobial activity and soybean growth. *Agric. Ecosystems Environ.,* 29:429–434.

Vera-Castello, J. C. and Ferrera-Cerrato, R., 1990. La Micorriza Vesículo-Arbuscular en Chinampas de San Gregorio Atlapulco, Xochimilco, México, Proceedings XXIIIth Mexican Congress of Soil Science, p. 151.

Vilariño, A. and Arines, J., 1992. The influence of aqueous extracts of burnt or heated soil on the activity of vesicular-arbuscular mycorrhizal fungi propagules. *Mycorrhiza,* 1:79–82.

Vilariño, A. and Arines, J., 1993. Changes in the development of *Acaulospora scrobiculata* in *Trifolium pratense* (red clover) roots and bulk substrate after plant burning. *Plant Soil,* 148:7–10.

Wacquant, J. P., Ouknider, M., and Jacquard, P., 1989. Evidence for a periodic excretion of nitrogen by roots of grass-legume associations. *Plant Soil,* 116:57–68.

Williams, P. G., 1985. Orchidaceous rhizoctonias in pot cultures of vesicular-arbuscular mycorrhizal fungi. *Can. J. Bot.,* 63:1329–1333.

Zuckerman, B. M., Dicklow, M. B., Coles, G. C., Garcia, R., and Marbán-Mendoza, N., 1989. Suppression of plant parasitic nematodes in the chinampa agricultural soils. *J. Chem. Ecol.,* 15:1947–1955.

# CHAPTER 6

# Role of Earthworms in Traditional and Improved Low-Input Agricultural Systems in West Africa

S. Hauser, B. Vanlauwe, D. O. Asawalam, and L. Norgrove

## INTRODUCTION

### Low-Input Agricultural Systems

Sub-Saharan Africa is the only part of the world where the per capita food production has declined in the last 20 years (IBRD, 1989; Ehui and Spencer, 1990, 1992). Increasing population density has led to an increase in demand for food. Farmers have responded by shortening regenerating fallow periods (Goldman, 1990). Land depreciation, indicated by incomplete restoration of soil fertility and decline in crop yields, is the result (OTA, 1984; Matlon and Spencer, 1984). Cultivation of an increasing proportion of land is thus required, causing a diminishing natural resource base (Ehui and Hertel, 1989; Ehui et al., 1990), as well as the destruction of the natural habitat for plant and animal species.

Small-scale farmers in West Africa have only scarce or no financial resources to purchase agricultural inputs; the few purchased are mainly used for cash crops such as cocoa and coffee. Due to infrastructural, economic, and soil-related problems of pesticide and fertilizer use, high-input, intensive agriculture as in developed countries is rarely practiced (Lavelle et al., 1992). Thus a large portion of arable land is still managed in the traditional way of "slash and burn," with its large land yet low capital and low labor requirements. Social (food supply) and environmental concerns over the continued clearing of forests have led to the development of alternatives to slash and burn.

Innovative systems should permit higher yields for a longer period of continuous cropping, yet should require low or no external inputs while increasing the sustainability of the land-use system. Two possibilities are alley cropping (Kang et al., 1984) and live mulch systems (Akobundu, 1980, 1984). In alley cropping, food crops are grown between hedgerows of trees or shrubs that are pruned during the cropping phase. A live-mulch system consists of a herbaceous legume species interspersed with food crops. During cropping, the herbaceous legumes are slashed. In both systems cutting the biomass reduces competition and provides soil-protecting mulch and nutrients. Slashing of live mulch also prevents climbing species from overgrowing and breaking the crop. Both systems are supposed to achieve higher nutrient recycling and use-efficiency through a multi-layered, deep-reaching root system and the use of phases in which food crops cannot be grown. A second aspect is the option of a controlled fallow and the immediate presence of a soil-regenerating species after cropping.

Both systems have been investigated over a number of years and permit higher yields (Mulongoy and Akobundu, 1985; Kang et al., 1990; Lathwell, 1990; Hauser and Kang, 1993; Kühne, 1993) and longer continuous cropping, yet reduced fertilizer inputs. Establishment of these systems requires little or no capital investment, but might increase the total demand for labor, cause seasonal shifts in labor allocation, and need some managerial skills. Farmer-participatory research on-farm has shown that alley cropping is a suitable option (Akonde et al., 1989; Getahun and Jama, 1989; Parera, 1989). However, neither system permits continuous cropping without declining yields and soil degradation, although the decline is slower than in traditional slash and burn (Van der Meersch, 1992; Hauser and Kang, 1993).

Past research has focused mainly on aboveground properties and performance of the vegetation, and very little information has been gathered on below-surface features of improved, as well as traditional, cropping systems (Lal, 1991). The latter is particularly true for soil biological activity, especially for the soil macrofauna, including earthworms (Brussaard et al., 1993).

## The Potential of Earthworm Activity

Soil-related constraints, such as low inherent fertility, usually limit crop production in the more humid areas of West Africa. Crop nutrition thus relies on biological processes that are mediated by teams of soil fauna in which earthworms play important roles. If soils are to be managed so that their biological capacity for nutrient cycling and maintenance of soil structure is retained, then more attention should be paid to the effect of cultivation and cropping practices on earthworms (Springett et al., 1992).

Earthworms' important role in soil profile development (Bouché, 1981), soil restoration, and maintenance of soil properties has been shown for a wide range of conditions (Edwards and Lofty, 1972; Satchell, 1983; Lee, 1985; Blanchart, 1992). However, earthworms are not primary producers, but transform and translocate soil, soil organic matter, and plant nutrients, so they depend on the vegetation and other organisms to provide food sources and favorable biophysical conditions. A large number of publications recently summarized by Lavelle (1994) describe the beneficial effects of earthworm activity and their casts on soil properties, plant growth, and ecosystem stability.

Earthworm activity has physical and biochemical consequences for agriculture. Earthworms burrow, improving macroporosity (Brussaard et al., 1990; Marinissen and Dexter, 1990) and infiltration properties (Ehlers, 1975; Douglas et al., 1980; Lal, 1987; Casenave and Valentine, 1989). While burrowing, they ingest large amounts of soil and plant residue. In Lamto, Cote d'Ivoire, Megascolecidae, and Eudrilidae species consume 6.7 g dry weight per 1 g individual per day. Of this, 99.9% is egested as casts (Lavelle, 1974) deposited at the surface, in burrows or in other macropores.

Casts usually contain more organic carbon, total nitrogen, and exchangeable cations than the surrounding topsoil (De Vleeschauwer and Lal, 1981; Lal and De Vleeschauwer, 1982; Mulongoy and Bedoret, 1989; Fragoso et al., 1993; Hauser, 1993). Casts also have higher microbial populations and enzyme activity than the ingested soil (Gorbenko et al., 1986; Tiwari et al., 1989; Barois et al., 1993; Tiwari and Mishra, 1993). There is some evidence that earthworms preferentially ingest smaller soil particles, so casts contain more clay and silt and less sand than the soil in which they live (Nye, 1955; Watanabe, 1975; Sharpley and Syers, 1976; Lavelle et al., 1992; Hauser, 1993).

Since agricultural production is usually accompanied by a major disturbance of the natural ecosystem, three basic questions need to be answered to assess the role of earthworms in sustainable, low-input agricultural systems:

1. Does earthworm activity make a significant contribution to the sustainability of natural ecosystems?
2. What are the key factors affecting their survival and activity?
3. Can management techniques be manipulated to maintain activity during phases of disturbance such as cropping?

Little or no information is available to answer all three questions for traditional and alternative cropping systems within one particular environment. This paper reports on a series of field experiments and investigations on earthworm activity in a subhumid and a humid tropical environment in West Africa.

## Surface Casting as an Index of Earthworm Activity

Earthworm activity includes burrowing, ingesting soil, transforming it, and exporting it as casts. Activity has previously been described by using biomass of surface litter removed (assuming earthworms are the only organisms doing this), volume and length of burrows excavated, and numbers or dry weight of casts. However, the majority of studies have quantified activity using number and/or biomass of worms expelled from the soil. This assumes that the volume of burrows excavated, soil ingested, and casts egested by a population of worms is proportional to its size. Yet recent research shows that these are significantly affected by food quality, soil moisture levels, and temperature (Martin and Lavelle, 1992; Kretzschmar and Bruchou, 1991).

The use of casting as an index of activity has a number of advantages. Casting is an actual expression of egestion which is correlated to ingestion in nutrient-poor soils. Sampling is nondestructive, allowing repeated measurements in the same field area over time. In contrast, assessing the volume of burrows excavated in the field is only possible by destructive measures. It is not easy to assess the quantity of subsurface casts or its significance in aggregate formation and soil structure (Lee, 1985), but those deposited at the surface are easy to quantify, have a more significant effect on soil structure and profile development (Bouché, 1981), and minimize the risk of soil losses through surface runoff and erosion (Hauser, 1990). Surface casting species are known among all the families of earthworms (Lee in Satchell, 1983). Syers et al. (1979) reported that surface cast production was correlated to removal of surface litter, thus confirming the strong link between surface casting and earthworm activity. Surface casting as an index is also suggested by Edwards and Lofty (1972).

## MATERIALS AND METHODS

Experiments and observations were conducted between 1990 and 1994 at IITA headquarters, Ibadan (7° 31′ N and 3° 54′ E), southwestern Nigeria, and at the IITA Humid Forest Station, Mbalmayo (3° 51′ N and 11° 27′ E), southern Cameroon. The annual rainfall at Ibadan is 1200 mm, with a bimodal distribution. Rains commence in April, followed by a short dry season during August, then recommence in September, and stop at the end of October. Soils are mainly Alfisols (Oxic Paleustalf) on the upper slopes and Entisols (Psammentic Ustortent) on the lower slopes and in valleys (Moormann et al., 1975). At Mbalmayo annual rainfall is 1520 mm, with a bimodal distribution. Rains commence in March and end in early July, followed by a short dry season of 6 to 8 weeks, then recommence in September, and stop at the end of November. The soil is classified as a clayey, kaolinitic, isohyperthermic, Typic Kandiudult (Hulugalle and Ndi, 1993). At both sites vegetation is humid, semi-deciduous, mature and young secondary forest. At both sites field experiments were conducted only on manually cleared land. At Mbalmayo casting activity was monitored in undisturbed secondary forest that had not been cultivated for at least 20 years. This was compared with activity in slashed-and-burned fields planted to an intercrop of maize, cassava, groundnut, and plantain. At Ibadan, forest and natural bush regrowth were compared with alley cropping using *Leucaena leucocephala, Senna siamea* or *Dactyladenia barteri* as hedgerow species and herbaceous legume live mulch using *Pueraria phaseoloides.* All data on earthworm casting activity were obtained using a continuous sampling method. Surface casts were collected once or twice per week from framed microplots. Casts were dried at 65°C after each sampling and analyzed after the end of the casting season.

## RESULTS AND DISCUSSION

### Methodological Aspects of Monitoring Earthworm Activity

The literature on earthworm activity in West Africa provides a wide range of data from various environments; however, there is no common methodology for calculating total annual soil ingestion and cast deposition. For example, if casting levels for sampling that does not cover the whole season are extrapolated, serious errors may occur because of pronounced phases of casting and no-casting (for Ibadan, see Figure 1). As a result, data are wide-ranging, although this may be caused by environmental conditions.

Sampling frequency is another critical issue. Fresh casts and casts that have not dried at least once are not very stable and can easily be destroyed by rain. A high sampling frequency is thus required to reduce the risk of underestimating casting. In an experiment where the impact of rain was reduced by 2-mm mesh screen and cotton cloth, cast recovery at weekly samplings was increased by 21.8% compared with plots receiving rain at full impact (Figure 2).

When comparing casting in a different ecosystems such as forests vs. cropped fields, the effect of altered raindrop size on casts must be considered. Although the amount of throughfall is lower in forests, the drop size and therefore the detachment capacity is higher (Evans, 1980; Lal, 1987), and a higher rate of cast disintegration in forests can be expected. Conversely, a live mulch with a multi-layered canopy close to the soil surface such as in *P. phaseoloides* live mulch might greatly reduce mechanical disintegration. Thus amounts of casts collected are probably lower than the amounts deposited. The potential errors increase with decreasing ground cover and increasing canopy height.

### Earthworm Species and Their Distribution

Earthworms are widespread in West Africa except where the mean annual rainfall is less than 800 to 1000 mm and the dry season exceeds 3 to 5 months (Lavelle, 1983). More than 28 genera are represented (Table 1). In Ibadan, the most frequently found species in descending order of importance are *Hyperiodrilus africanus,* and *Eudrilus eugeniae. H. africanus* is a surface casting species reported not to feed on litter at the surface (Madge, 1965). In the forest and in newly cleared sites a large species of up to 30 cm length was found.

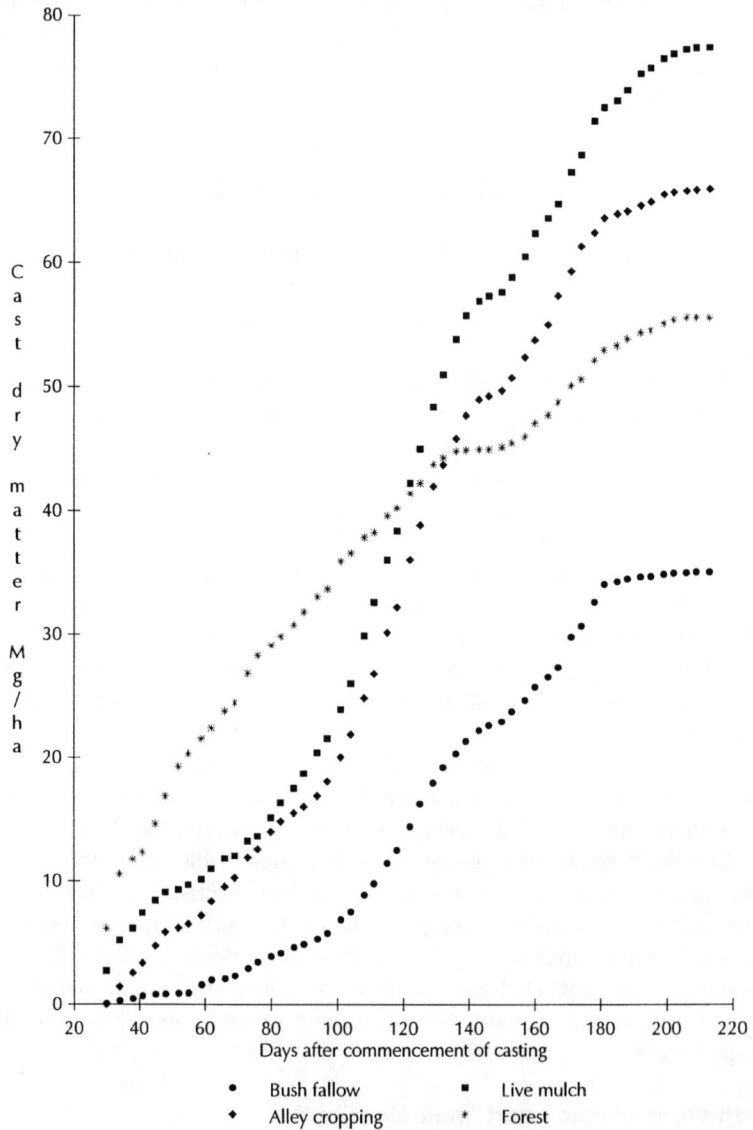

**Figure 1**  Annual cumulative amount of casts deposited at the soil surface — Ibadan, Nigeria, 1992.

This species is yet to be identified. It is not very abundant and is not found in older fields.

The two dominant species are not uniformly distributed between cropping systems. In newly cleared fields, both are abundant. However, with an increasing number of years of cropping and under conditions exposing the soil surface

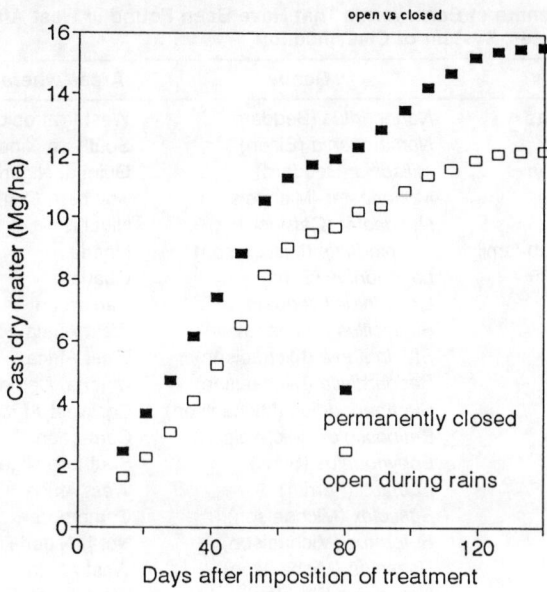

**Figure 2**  Cumulative amount of casts recovered from plots permanently covered with screen or cotton cloth vs. plots kept open during rains.

for a long time to direct radiation and full impact of rain, *H. africanus* disappears, while *E. eugeniae* becomes dominant. Yet, where mulch is provided and a cover crop is grown, or near trees in alley-cropping systems, both species remain abundant and *H. africanus* continues to be dominant.

Two basic types of surface casts are found: pipe-shaped casts with a vertical hole running through the length, sealed at the top; and those composed of fine granular pellets stuck together. Madge (1969) found that these are species specific; the former egested by *H. africanus* and the latter by *E. eugeniae*. At Mbalmayo casts similar to the *H. africanus* casts but larger in diameter are found. However, the dominant species producing these casts is yet to be identified.

## Earthworm Activity in Relatively Undisturbed Environments

Kollmannsperger (1956) reports of 25 to 30 Mg ha$^{-1}$ surface casts annually in the Cameroonian mountain savannah. Madge (1969) calculated an annual surface cast production of 30 to 240 Mg ha$^{-1}$ in grassland in southwestern Nigeria. Lal and Cummings (1979) estimated cast deposition of 328 Mg ha$^{-1}$ yr$^{-1}$ in a forest in southwestern Nigeria. Lavelle (1978) reports of 278 Mg ha$^{-1}$ yr$^{-1}$ of surface casts in grass savannah, while Beauge (1912) found 268 Mg ha$^{-1}$ yr$^{-1}$ in grassland in Sudan Gezira (adapted in Lee, 1985). Very little is known about earthworms' contribution to organic matter turnover and nutrient cycling or their impact on soil texture and structural stability in their natural undisturbed environments.

**Table 1    Genera of Earthworms That Have Been Found in West Africa Using Gates' System of Classification**

| Family | Genus | Areas where identified |
|---|---|---|
| Ocnerodrilidae | *Nannodrilus* (Beddard) | Western tropical Africa |
| | *Nematogenia* (Eisen) | Southern Nigeria, Liberia |
| Octochaetidae | *Millsonia* (Beddard) | Guinea, Nigeria |
| | *Monogaster* (Michaelson) | Southern Cameroon |
| | *Neogaster* (Cernosvitoz) | Nigeria |
| Eudrilidae sub-family | *Chuniodrilus* (Michaelson) | Liberia |
| Parendrilinae | *Legondrilus* (Sims) | Ghana |
| | *Libyodrilus* (Beddard) | Cameroon |
| | *Scolecillus* (Omodeo) | West Africa |
| | *Stuhlmannia* (Michaelson) | West Africa |
| | *Beddardiella* (Michaelson) | Nigeria, Cameroon |
| | *Buettneriodrilus* (Michaelson) | Eq. West Africa |
| | *Eminoscolex* (Michaelson) | Cameroon |
| | *Ephyriodrilus* (Sims) | Southern Nigeria |
| | *Eudrilus* (Perrier) | West Africa |
| | *Euscolex* (Michaelson) | Cameroon |
| | *Eutoretus* (Michaelson) | North Nigeria |
| | *Haaseina* (Michaelson) | West Africa |
| | *Heliodrilus* (Beddard) | West Africa |
| | *Hippopera* (Taylor) | West Africa |
| | *Hyperiodrilus* (Beddard) | Togo, S. Nigeria |
| | *Iridodrilus* (Beddard) | Cameroon |
| | *Kaffania* (Michaelson) | Cameroon |
| | *Keffia* (Clausen) | West Africa |
| | *Parascolex* (Michaelson) | Cameroon, Togo |
| | *Rosadrilus* (Cognetti) | Cameroon |
| | *Teleutoreutus* (Michaelson) | West Africa |
| Microhaetidae | *Alma* (Grube) | Cameroon, Nigeria, Togo |

Adapted from Edwards, C. A. and Lofty, J. R., 1977. *Biology of Earthworms,* 2nd ed., Chapman and Hall, London, p. 333.

In Ibadan, the forest is the least disturbed ecosystem followed by fallow land of decreasing fallow length, representing systems with increasing impact of human activity. An experiment was set up in Ibadan to compare surface casting in the forest with various 3-year fallow systems: bush regrowth, *Leucaena leucocephala* fallow, and *Pueraria phaseoloides* fallow. Although locally and in very small areas, casting of more than 300 Mg ha$^{-1}$ was observed, the average annual cast deposition was 38.5 Mg ha$^{-1}$ in the forest and 80 Mg ha$^{-1}$ in the bush regrowth (Table 2). In the *L. leucocephala* fallow, casting was 33% higher than in the forest. In the *P. phaseoloides* live mulch reported here, invasion by carnivorous ants drastically reduced the earthworm population at the start of the season. This indicates that the least disturbed system does not provide the best conditions for maximum surface casting, organic matter turnover, and nitrogen cycling. This might result from characteristics of the soil moisture regime under forests. During dry phases soil water tension in the top 50 cm increased faster under forests than under the other treatments

**Table 2  Annual Cast Deposition and Amount and Concentration of Organic Carbon and Total Nitrogen from Forests and 3-Year-Old Fallows on Alfisols, Ibadan, 1991**

|  | Casts (Mg ha⁻¹) | Org. C (kg/ha) | Ttl. N (kg/ha) | Org. C (%) | Ttl. N (%) |
|---|---|---|---|---|---|
| Forests | 38.5 | 2619.5 | 213.0 | 6.74 | 0.55 |
| Bush regrowth | 80.2 | 3699.2 | 360.7 | 4.52 | 0.45 |
| *L. leucephala* regrowth | 51.1 | 2924.5 | 243.3 | 5.51 | 0.46 |
| *P. phaseoloides* regrowth | 19.8[a] | 1376.7 | 125.0 | 6.92 | 0.62 |

[a] Earthworm population drastically reduced by carnivorous ants.

(Figure 3). This might have caused an earlier retreat of worms to deeper layers and, consequently, less activity near the surface. Martin and Lavelle (1992) showed in simulations that soil water content is a key factor in earthworms' vertical movements.

The amount of organic carbon in casts represents 5.0 to 11.6% of the total organic carbon, while the total nitrogen in casts ranges between 5.3 and 12.9% of the total in the top 0 to 15 cm of the soil profile (Table 3). Thus earthworm casting activity involves a considerable proportion of the soil carbon and nitrogen pool in two of the fallow systems. The comparatively low proportions

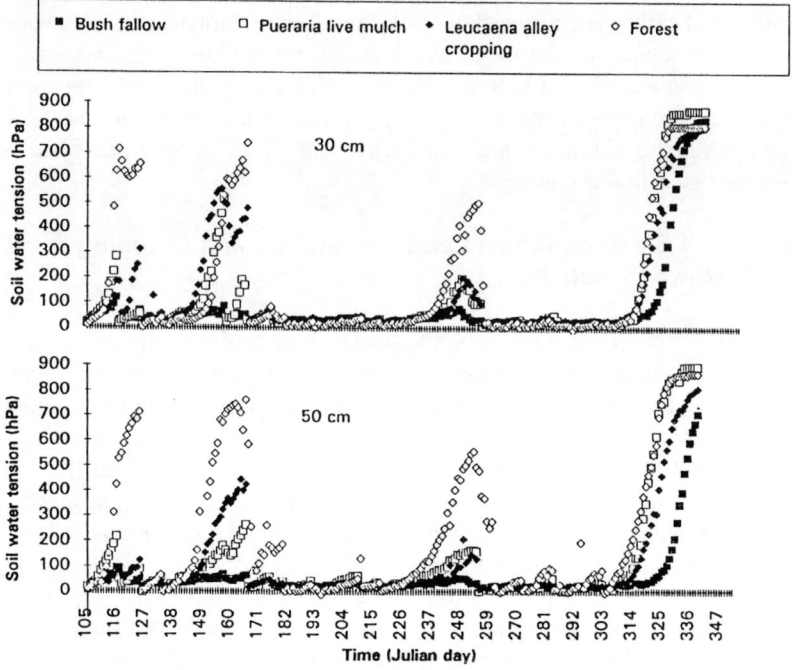

**Figure 3**  Soil water tension at 30 and 50 cm depth under forest and fallows — Alfisol, Ibadan, Nigeria, 1991.

Table 3    Amounts of Organic Carbon and Total Nitrogen in the 0 to 15 cm Topsoil and Proportion Contained in Earthworm Casts — Alfisol, Ibadan, 1991

|  | Soil organic C (Mg ha$^{-1}$) | Soil total N (kg/ha) | Percentage in casts | |
|---|---|---|---|---|
|  |  |  | Org. C | Ttl. N |
| Forests | 54.6 | 3192.2 | 5.74 | 6.67 |
| Bush regrowth | 32.5 | 2803.7 | 11.38 | 12.86 |
| L. leucephala regrowth | 25.3 | 2099.1 | 11.58 | 11.59 |
| P. phaseoloides regrowth | 27.4 | 2357.9 | 5.03[a] | 5.30[a] |

[a] Earthworm population drastically reduced by ants.

in the casts from forest soil relate to the very high amounts of organic carbon and nitrogen in the forest soil. The low values in *P. phaseoloides* live mulch are due to the low casting activity, since the chemical properties of the casts would have resulted in a similar or higher proportion if the amount of casts had been comparable with that in the forest soil.

Relating the amounts of carbon and nitrogen incorporated in casts to the soil carbon and nitrogen pool does not reflect the importance of earthworm activity in other processes of nutrient recycling and organic matter turnover. However, the above figures indicate that earthworm casts are more important after forest clearance since they contain a greater proportion of the soil nutrients as soil carbon and nitrogen pools decline. The performance of earthworms should be related to processes like biomass production of the vegetation, nutrient accumulation in biomass, decomposition of biomass, and release of nutrients. Such investigations were not possible in the forest and during the regrowth phase of fallows, but data are available from cropped fields and are reported later in this chapter.

## Impact of Slash-and-Burn Land Preparation and Cropping on Earthworm Activity

Human activity can change biophysical conditions drastically as is the case when forests or fallows are cleared to grow food crops (Critchley et al., 1979; Lal, 1986). Little is known about the immediate impact on earthworm activity of converting forest or fallow into arable land.

In 1992 the three fallows mentioned above were cleared and the bush regrowth, as well as the *P. phaseoloides,* completely burned. In *L. leucocephala* alley cropping, only the understorey was burned, but the *L. leucocephala* was cut after the burn and left on the plots until the leaves were shed. Wood was then removed, and all plots were planted to maize/cassava intercrop. In the same experiment were permanently cropped plots under the same fallow managements. They had been cropped for the previous 3 years and were entering the fourth year of cropping.

Table 4   Casting Activity, Organic Matter Accumulation, and Nutrient
Recycling in Newly Cleared (New) Vs. Permanently Cropped (Perm.)
Fallow Management Systems and Secondary Forest — Alfisol,
Ibadan, Nigeria, 1992

| | Casts (Mg ha⁻¹) | Org. C (kg/ha) | Ttl. N (kg/ha) | Exch. Ca (kg/ha) | Exch. Mg (kg/ha) |
|---|---|---|---|---|---|
| Forest | 75.0 | 5395 | 407.1 | 283.9 | 47.8 |
| Bush fallow perm. | 60.0 | 2622 | 200.0 | 178.3 | 24.0 |
| Bush fallow new | 28.8 | 1679 | 94.1 | 72.3 | 14.5 |
| Alley cropping perm. | 91.6 | 4501 | 322.4 | 250.0 | 39.8 |
| Alley cropping new | 59.4 | 3170 | 246.1 | 203.0 | 26.2 |
| *Pueraria* mulch perm. | 86.3 | 4554 | 351.6 | 321.2 | 27.3 |
| *Pueraria* mulch new | 55.2 | 2764 | 212.4 | 171.4 | 43.3 |

Casting activity and the amount of organic carbon and nutrients in casts were higher in the permanently cropped plots than in newly cleared plots for all management systems (Table 4). Of the permanently cropped treatments, the two improved fallow management systems had the highest casting, exceeding that in the forest. Chemical properties of casts from the forest were enriched in nutrients and organic carbon compared with casts from cropped fields (Table 5). Only exchangeable magnesium was higher in the permanently cropped *P. phaseoloides* live mulch system.

Lower casting activity in newly cleared compared with permanently cropped plots is an unexpected result. It indicates that drastic environmental change severely disrupts earthworms. Deep infiltration through macropores of rain, high in pH from dissolving ash, apparently has a detrimental effect on casting activity. The negative impact of ash on casting was confirmed in separate experiments (Asawalam, unpublished). However, the heat from burning could not have had an effect since the burning was performed before the worms appeared in the surface soil.

Exposure of the soil surface to direct radiation during clearance may also be significant. The importance of ground cover or shading for high casting

Table 5   Chemical Properties of Earthworm Casts from Newly
Cleared (New) Vs. Permanently Cropped (Perm.) Fallow
Management Systems and Secondary Forests — Alfisol,
Ibadan, Nigeria, 1992

| | Org. C (%) | Ttl. N (%) | Exch. Ca (cmol[+]/kg) | Exch. Mg (cmol[+]/kg) |
|---|---|---|---|---|
| Forests | 7.05 | 0.545 | 19.0 | 5.2 |
| Bush fallow perm. | 4.39 | 0.336 | 14.6 | 3.3 |
| Bush fallow new | 5.57 | 0.373 | 13.9 | 4.2 |
| Alley cropping perm. | 4.83 | 0.355 | 13.5 | 3.3 |
| Alley cropping new | 5.35 | 0.435 | 17.5 | 3.6 |
| *Pueraria* mulch perm. | 5.23 | 0.405 | 18.7 | 2.6 |
| *Pueraria* mulch new | 4.90 | 0.377 | 15.8 | 6.4 |

activity has been shown by Franzen (1986) and Hauser (1993). Weeds, crop residues, slashed *P. phaseoloides,* and *L. leucocephala* prunings provided ground cover in the early phases of crop development. The possible increase in food supply from decomposing roots apparently does not compensate for the negative impact of exposure to the sun. This agrees with Hauser (1993) who demonstrated that shade is more important than food supply.

In Mbalmayo, casting was severely reduced in the cropped fields. Mean annual casting was 2.82 Mg ha$^{-1}$, while in the adjacent forest it was 9.3 Mg ha$^{-1}$. In plots maintained bare on the field periphery only 0.87 Mg ha$^{-1}$ of casts were recorded.

## Performance of Earthworms in Permanently Cropped Fields

Casting activity and nutrient cycling in cropped fields can exceed that in forests (Table 4). Management practices such as burning vs. mulching apparently have a major impact. Over time, however, activity declines in all cropping systems (Figure 4). The regression suggests that casting is initially higher under alley cropping than in the traditional system without trees. Unfortunately, there are no data available on casting activity in the first 3 years after clearing without the impact of burning. Thus it might be that in traditional systems a more rapid decline in casting occurs in the first few years, while it declines more steadily in alley cropping. As casting activity was higher in the alley cropping treatment (cleared from the forest 3 years before) than in the

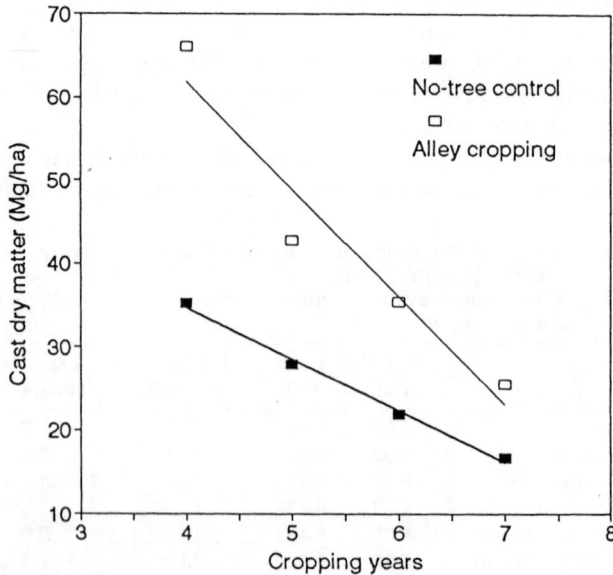

**Figure 4**  Annual earthworm casts deposition at the surface as a function of cropping years.

Table 6   Dry Matter Production, Nitrogen Concentration and Accumulation, Amount of Residues at the Onset of Rains, and Root Density During the Rainy Season of *P. phaseoloides* Live Mulch and *L. leucocephala* Alley Cropping on Alfisol, Ibadan, Nigeria

| Species | Biomass (Mg ha⁻¹) | Nitrogen (%) | Nitrogen (kg/ha) | Residues (Mg ha⁻¹) | Root[b] density/1500 cm² |
|---|---|---|---|---|---|
| *Pueraria* | 5.62 | 2.02 | 113.6 | 4.15 | 262.0 |
| *Leucaena*[a] | 4.50 | 4.19 | 188.5 | 0.51 | 48.2 |

[a] Excluding wood.
[b] Counted on surface of a trench, 0 to 150 cm depth.

forest (Table 4), there was a net increase in casting during the 3 years of treatment.

To use earthworm activity for sustaining soil fertility, factors that stimulate casting activity in improved fallow and cropping systems need to be determined. Both *P. phaseoloides* live mulch and alley cropping provide additional biomass to the production of the weeds and food crops. Average *P. phaseoloides* biomass production on an Alfisol was 5.62 Mg ha⁻¹ (Table 6) with a maximum of 9.0 Mg ha⁻¹. Maximum dry matter production was attained relatively late in the year (end of November), so that a large amount of biomass was retained throughout the dry season until the onset of rains. Alley cropping with *Leucaena leucocephala* on an Alfisol produced on average, 8.63 Mg ha⁻¹ dry matter. This amount comprised 52% leaves and small twigs and 48% woody stems. Nutrient release and decomposition of the two materials are probably quite different because *L. leucocephala* leaves had twice as high a nitrogen concentration of *P. phaseoloides*.

On the more fertile Alfisols, the differences in biomass production and litter retention did not cause a pronounced difference in casting activity between *P. phaseoloides* live mulch and alley cropping. The 30% lower casting activity in bush fallow regrowth might be caused by the lack of litter and pruning inputs. On the less fertile sandy Entisols, however, the positive effect of the more persistent litter of *P. phaseoloides* and the possibly higher fine-root turnover permitted casting activity almost twice that in alley cropping. Lack of sufficient ground cover and biomass input in the bush fallow regrowth led to a 75 to 86% reduction in casting, as compared with alley cropping and *P. phaseoloides* live mulch, respectively (Table 7).

Table 7   Amounts of Casts, Organic Carbon, and Total Nitrogen Deposited at the Soil Surface — Entisol, Ibadan, 1992

| | Casts (Mg ha⁻¹) | Organic C (kg/ha) | Total N (kg/ha) |
|---|---|---|---|
| Forests | 36.5 | 1805.4 | 131.5 |
| Bush fallow system | 10.4 | 412.6 | 27.5 |
| *Leucaena* alley cropping | 40.6 | 1903.6 | 136.6 |
| *Pueraria* live mulch | 60.0 | 2896.9 | 221.5 |

On the comparatively poor Ultisol at Mbalmayo no surface casting was observed in the third year of alley cropping using *Senna spectabilis, Dactyladenia barteri,* or *Flemingia macrophylla* as hedgerow trees. The only crop maintaining surface casting was plantain, but it was restricted to the close vicinity of the corm, which is heavily mulched with the residues of harvested plants.

The results indicate that cropping systems targeted at more sustainable use of the soil resource can only be developed if the interdependencies between soil type and the most compatible vegetation are known and considered. They also show that earthworms react more sensitively to disturbance on less fertile soils.

## Spatial Heterogeneity of Earthworm Activity in Alley Cropping

In alley cropping the supply of food through aboveground prunings is equal at all positions in the system, while the persistence of shade varies over time and distance from hedgerows. Only in the immediate vicinity of the hedgerows is the soil shaded all year round, and it was there that the highest casting activity was found in all alley cropping experiments (Table 8). In all *L. leucocephala* alley cropping systems older than 1 year, casting significantly declined toward the middle of the interrow space. When using *Senna siamea* or *Dactyladenia barteri,* producing a more recalcitrant litter, the decline was less pronounced but still significant in *Senna siamea.* Only on nondegraded soil was casting activity in the middle of the interrow space significantly higher than in a system without trees.

In alley cropping the biological degradation process occurs at two contrasting locations with two different rates. Casting activity did not differ between 4-, 5-, and 6-year-old alley cropping systems under the hedgerows.

Table 8   Annual Casting Activity in Alley Cropping Under the Hedgerow (Row) and in the Interrow Space (Middle) as Compared with Casting in a No-Tree Control

| Hedgerow species | Cropping years | Casts (Mg ha$^{-1}$) | | | |
|---|---|---|---|---|---|
| | | Row | Middle | No-tree | LSD 0.05 |
| *Leucaena leucocephala* | 1[a] | 63.9 | 56.0 | 27.4 | 18.7 |
| *Leucaena leucocephala* | 4[a] | 113.9 | 60.6 | 35.2 | 15.2 |
| *Leucaena leucocephala* | 5[b] | 116.8 | 24.3 | 27.8 | 7.8 |
| *Leucaena leucocephala* | 6[c] | 49.9 | 14.1 | 15.6 | 23.7 |
| *Leucaena leucocephala* | 6[b] | 112.8 | 23.1 | 21.8 | 46.2 |
| *Dactyladenia barteri* | 6[b] | 50.5 | 36.6 | 21.8 | 16.7 |
| *Leucaena leucocephala* | 7[c] | 93.1 | 12.7 | 16.6 | 47.6 |
| *Senna siamea* | 7[c] | 77.2 | 19.4 | 16.6 | 30.8 |

[a] Nondegraded.
[b] Moderately degraded.
[c] Severely degraded soil when alley cropping was implemented.

Table 9   Total Amounts of Casts Deposited at the Soil Surface in Alley Cropping and Amounts of Organic Carbon and Nutrients in Alley Cropping and No-Tree Control

| | | Alley cropping | | | No-tree | |
| Hedgerow species | Cropping years | Casts (Mg ha⁻¹) | Org. C (kg/ha) | Ttl. N (kg/ha) | Org. C (kg/ha) | Ttl. N (kg/ha) |
| --- | --- | --- | --- | --- | --- | --- |
| Leucaena leucocephala | 1[a] | 53.5 | 2527 | 200 | 1420 | 89 |
| Leucaena leucocephala | 4[a] | 66.1 | 3202[d] | 230 | 1518 | 114 |
| Leucaena leucocephal | 5[b] | 42.8 | 1510[d] | 138[d] | 490 | 46 |
| Leucaena leucocephala | 6[c] | 23.5 | 857 | 85 | 501 | 37 |
| Leucaena leucocephala | 6[b] | 35.3 | 1743[d] | 138 | 625 | 48 |
| Dactyladenia barteri | 6[b] | 36.4 | 1555[d] | 89 | 625 | 48 |
| Leucaena leucocephala | 7[c] | 25.5 | 1160 | 89 | 411 | 28 |
| Senna siamea | 7[c] | 31.4 | 1371 | 96 | 411 | 28 |

[a] Nondegraded.
[b] Moderately degraded.
[c] Severely degraded soil when alley cropping was implemented.
[d] Significantly different ($p$, 0.05) from respective values in no-tree control.

Under moderately and severely degraded soil conditions the application of prunings in the interrow space had no significant effect. Since the immediate vicinity of the hedgerow is only a small portion of the alley cropping system, the weighted casting activity of the whole system was closer to the amounts found in the interrow space (compare Tables 8 and 9). Chemical properties of casts did not significantly differ between positions in alley cropping, so that organic carbon and total nitrogen were distributed as heterogeneously as the amounts of casts. Weighted-average deposition of organic carbon and total nitrogen was generally higher in alley cropping, but the differences between alley cropping and the no-tree control were significant in only a few cases. Thus alley cropping does not maintain high casting activity over time. Narrowing interrow distances or the introduction of a cover crop as suggested by Hauser (1993) might be required to reduce soil degradation between hedgerows, but might reduce crop yield to unacceptably low levels.

## Contribution of Earthworm Activity to Soil Organic Matter and Nitrogen Turnover

The organic carbon and total nitrogen concentrations in the casts of *H. africanus* generally exceeded those in the corresponding topsoil. The increase of organic carbon in casts is relatively higher on soils of lower carbon content, and increments decrease with increasing soil carbon content (Figure 5). *H. africanus* has a particularly high potential to improve soil properties of poor soils, as long as other physical (shade) conditions permit a high level of activity. Madge (1965) reports that *H. africanus* does not feed on surface litter, so it is important to determine whether the increased organic carbon in casts is caused by concentration of soil organic matter through preferential uptake or whether *H. africanus* incorporates decomposers and decomposing fresh inputs to form new soil organic matter.

Ideally, earthworm activity should be evaluated in comparison with all other processes contributing to the maintenance of favorable soil properties. The primary and thus most important process in this respect is the biomass production of the vegetation and its nutrient accumulation (including nitrogen fixation). Most other factors depend on this primary production. It is very difficult to quantitatively determine and separate the turnover rates for organic carbon and nitrogen of all individual processes involved in a particular eco-system. We shall attempt to show the relationship between biomass production and nutrient accumulation of the vegetation and the turnover or incorporation of organic and total nitrogen into earthworm casts, here called the apparent incorporation rate (*AIR*), estimated by Equations 1 and 2.

$$AIR_{org\ C} = \text{org C in casts/org C in biomass} \cdot 100 \qquad (1)$$

$$AIR_{ttl\ N} = \text{total N in casts/total N in biomass} \cdot 100 \qquad (2)$$

Relating the measured amounts of nitrogen and organic carbon in the annual cast production to those in the aboveground biomass production or surface application of hedgerow prunings shows that with increasing the length of cropping and thus soil degradation the apparent incorporation rate decreases (Table 10). On nondegraded soil, apparent incorporation rates were generally above 100%. Under *P. phaseoloides* live mulch, more than three times more nitrogen was accumulated in casts than was determined in the aboveground *P. phaseoloides* biomass.

**Table 10  Apparent and Corrected Incorporation Rate of Total Nitrogen and Organic Carbon of Aboveground Organic Inputs Through Vegetation in Earthworm Casts**

| | | Incorporation rate | | | |
| | | Apparent | | Corrected | |
| Cropping system | Cropping years | Ttl. N | Org. C | Ttl. N | Org. C |
|---|---|---|---|---|---|
| *Leucaena* alley cropping | 1[a] | 197.1 | 248.4 | 142.8 | 182.1 |
| *Pueraria* live mulch | 1[a] | 186.3 | 109.2 | 67.0 | 70.9 |
| *Leucaena* alley cropping | 4[a] | 170.9 | 222.2 | 115.2 | 161.1 |
| *Pueraria* live mulch | 4[a] | 309.5 | 180.1 | 207.0 | 124.1 |
| *Leucaena* alley cropping | 5[b] | 49.1 | 47.8 | 29.9 | 32.9 |
| *Leucaena* alley cropping | 6[b] | 66.6 | 72.5 | 49.6 | 57.0 |
| *Dactyladenia* alley cropping | 6[b] | 74.4 | 50.8 | 55.4 | 39.4 |
| *Leucaena* alley cropping | 7[c] | 70.9 | 79.7 | 55.4 | 61.7 |
| *Senna* alley cropping | 7[c] | 54.0 | 52.8 | 40.6 | 39.4 |

[a] Nondegraded.
[b] Moderately degraded.
[c] Severely degraded soil.

The apparent incorporation rate can only indicate the potential for carbon and nitrogen retention in casts. To get a more accurate estimate of the actual incorporation of carbon and nitrogen, it is necessary to consider the portion of C and N taken up from the soil. For *H. africanus* no information is available onto what extent the worm ingests soil-borne carbon and nitrogen vs. carbon and nitrogen from decomposing dead or applied fresh material. For simplicity, it may be assumed that the earthworms take up soil at its original C and N concentration and, additionally, ingest decomposers and decomposing material originating from litter, root turnover, and applied fresh materials, which are not considered soil organic matter. The difference between cast and soil organic carbon and total nitrogen concentrations would be the portion obtained from these new organic inputs, or would be "nonsoil-borne." In combination with the amount of casts and C and N contents in the biomass the corrected incorporation rate (*CIR*) is estimated by Equations 3 and 4.

$$CIR_{org\ C} = [(\%\ \text{org C in casts} - \%\ \text{org C in soil}) \cdot \text{cast dry matter}]/\text{org C in biomass} \cdot 100 \tag{3}$$

$$CIR_{ttl\ N} = [(\%\ \text{ttl N in casts} - \%\ \text{ttl N in soil}) \cdot \text{cast dry matter}]/\text{ttl N in biomass} \cdot 100 \tag{4}$$

The proportion of nonsoil organic carbon in casts ranged from 69 to 79%, the proportion of nonsoil total nitrogen from 61 to 78%. There was a tendency toward higher proportions of nonsoil carbon and nitrogen with increasing soil degradation and length of cropping (compare with Figure 5).

On nondegraded soil cleared from forest, even after 4 years still more carbon and nitrogen were available for uptake by worms than were provided by aboveground organic inputs (Table 10). This might indicate that there is still a high amount of subsurface material decomposing and being taken up by earthworms. Incorporation of nitrogen in casts was highest in *P. phaseoloides* live mulch, which might be an indication of a high root and nodule turnover. On degraded soils, earthworm activity can incorporate one- to two-thirds of the carbon from organic inputs. The incorporation of nitrogen is slightly lower. The method used here cannot identify the actual sources of carbon and nitrogen. Investigations on the food source of *H. africanus* have only been of a qualitative nature (Madge, 1969) and did not distinguish between soil organic matter and decomposing new inputs.

In an experiment where [15]N-labeled *L. leucocephala* and *D. barteri* prunings were applied, the average percentage of nitrogen in casts derived from *L. leucocephala* was 14.2% (4.6 to 18.2%), while the average percentage of nitrogen derived from *D. barteri* was 5.5%, ranging from 1.5 to 9.0%. These figures were obtained from a 6-year-old alley cropping experiment and are therefore of limited representation. However, it appears that *H. africanus* draws predominantly on the more decomposed soil resource pool, rather than on the

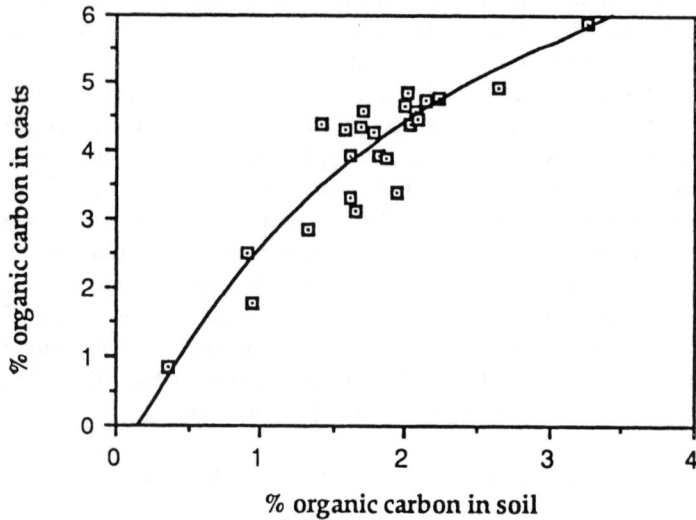

**Figure 5**    Relation between organic carbon concentration in the topsoil (0 to 5 cm) and in earthworm casts.

labeled fresh materials. In the case of *D. barteri* it is also necessary to acknowledge that the prunings are very recalcitrant, so that low relative uptake might rather be due to low release and availability. At 31 days after application of $^{15}N$-labeled prunings, total recovery of $^{15}N$ in casts was 22.5 and 9.2% from *L. leucocephala* and *D. barteri,* respectively. These figures show that earthworms ingest decomposing, newly applied materials or decomposers associated with them, and that incorporation of nitrogen into earthworm casts may compete in a cropping system with crops requiring high supplies of nitrogen.

## Earthworm Casts and Plant Growth

Earthworms can have positive effects on plant growth. Spain et al. (1992) showed that higher yields were associated with increased N uptake. Improved soil physical properties (Lal, 1988; Lal and Akinremi, 1983), as a consequence of burrowing and casting, may also contribute to enhanced crop performance. However, earthworms do not have positive effects on plant growth in all cases (Pashanasi et al., 1992).

Deposition of casts at the soil surface can generate a new soil layer of up to 0.55 cm thick within a year (Hauser, 1994). This layer has high concentrations of organic carbon and nutrients. Dry casts also have a high resistance to mechanical disintegration by raindrops. Thus a qualitative aspect of casting activity has to be considered. Experiments with *H. africanus* casts at Ibadan showed that casts withstood four times more rain events than soil aggregates before they started to disintegrate (Asawalam, unpublished).

In pot experiments, applying *H. africanus* casts to the soil surface or mixing ground casts into the soil, both increased maize growth and nitrogen uptake, as compared to maize in soil alone. Furthermore, cast application led to higher maize production in subsequent cropping phases, whereas area application did not (Mulongoy, 1990). In other experiments with *H. africanus* casts mixed with various amounts of topsoil, a linear increase of maize yield with an increasing proportion of casts in the mixture was shown (Asawalam, unpublished). Although the amount of nitrogen applied per pot tripled, the yield of maize increased by only 50%. Similar results were obtained at Mbalmayo using the same experimental approach. Nitrogen concentration in casts at Mbalmayo was only 58% higher than in the soil; however, the maize yield doubled in the pure casts compared with soil without casts. On the rather acid and poor Ultisol even small increases in nitrogen have a significant effect. Nitrogen availability from casts at Mbalmayo is apparently higher than from soil. Since this contradicts the results obtained at Ibadan and of other authors, it requires further investigation. Availability of nitrogen from casts is lower than from the topsoil at Ibadan. Thus a chemical or physical resistance to rapid mineralization and subsequent losses can be concluded. A stabilization of organic carbon by passage through earthworms has also been shown by Shaw and Pawluk (1986) and Lavelle et al. (1992). Since *H. africanus* casts contain up to four times more total nitrogen than the topsoil, an increased supply of nitrogen is possible even at lower release rates from the casts.

## CONCLUSION

Earthworms play an important role in low-input agricultural systems. Their casting activity involves up to 11.6% of the organic carbon and 12.9% of the total nitrogen of the 0 to 15 cm topsoil in undisturbed or recovering systems. Earthworms are very sensitive to changes in the ecosystem, expressed by strongly reduced surface casting activity. In improved cropping systems earthworm activity can exceed that in the forest. This increase is due to the maintained groundcover and its reduced water consumption compared with forests. The negative impact of traditional cropping (lack of organic matter input through vegetation management) on casting becomes more pronounced as soil fertility decreases, while permanent groundcover becomes more important. The benefit of earthworm activity, apart from the effects on soil physical properties, is its concentration and deposition of large amounts of organic carbon and total nitrogen at the surface. Resources are placed at a location and in a form where (and in which) they are least likely to be lost. A firm conclusion on the effect of casts on plant nutrition is not yet possible.

In sustainable agricultural production systems the resource-conserving aspect of earthworm activity might be the more important one. *H. africanus* is active over a broad range of soil qualities and shows that it has the potential to improve or mediate the buildup of favorable soil properties. Thus, in accordance with Lavelle et al. (1992), results from Ibadan show that earthworms are not merely a consequence of high soil fertility, but that they contribute to its buildup and maintenance.

## ACKNOWLEDGMENTS

These investigations were partially funded by the German Agency for Technical Cooperation (GTZ). The authors wish to thank Miss Charity Nnaji for typing and designing the graphics. We would also like to thank Dr. K. Vielhauer and the Honorary Consul Mr. H. Nau for their reliable technical support.

## REFERENCES

Akobundu, I. O., 1980. *Live Mulch: A New Approach to Weed Control and Crop Production in the Tropics,* Proceedings 1980 Brit. Crop Protection Conference — Weed, pp. 377–382.

Akobundu, I. O., 1984. *Advances in Live Mulch Crop Production in the Tropics,* Proceedings 1984 Western Society of Weed Science, 37:51–55.

Akonde, T. P., Lame, B., and Kummerer, E., 1989. Adoption of alley cropping in the Province of Atlantique, Benin, in *Alley Farming in the Humid and Subhumid Tropics,* Kang, B. T. and Reynolds, L., Eds., Proceedings of an international workshop held at Ibadan, Nigeria, March 10–14, 1986, IDRC, Ottawa, Canada, pp. 141–142.

Barois, I., Villemin, G., Lavelle, P., and Toutain, F., 1993. Transformation of the soil structure through *Pontoscolex corethrurus* (Oligochaeta) intestinal tract. *Geoderma,* 56:57–66.

Blanchart, E., 1992. Restoration by earthworms (Megascolecidae) of the macroaggregate structure of a destructured savanna soil under field conditions. *Soil Biol. Biochem.,* 24:1587–1594.

Bouché, M. B., 1981. Contribution des lombiciens aux migrations d'élements dans les sols tempérés. *Coll. CNRS,* 202:145–153.

Brussaard, L., Coleman, D. C., Crossley, D. A., Didden, W. A. M., Hendrix, P. F., and Marinissen, J. C. Y., 1990. Impacts of earthworms on soil aggregate stability. *Trans. 14th ICSS,* 3:100–103.

Brussaard, L., Hauser, S., and Tian, G., 1993. Soil faunal activity in relation to the sustainability of agricultural systems in the humid tropics, in *Soil Organic Matter Dynamics and Sustainability of Tropical Agriculture,* Mulongoy, K. and Merckx, R., Eds., John Wiley & Sons, Chichester, pp. 241–256.

Casenave, A. and Valentin, C., 1989. *Les Etats de Surface de la Zones Sahelienne. Influence sur l'Infiltration,* ORSTOM, Paris.

Chritchley, B. R., Cook, A. G., Critchley, U., Perfect, T. J., Russel-Smith, A., and Yeadon, R., 1979. Effects of bush clearing and soil cultivation on the invertebrate fauna of a forest soil in the humid tropics. *Pedobiologia,* 19:425–438.

De Vleeschauwer, D. and Lal, R., 1981. Properties of worm casts under secondary tropical forest regrowth. *Soil Sci.,* 132:175–181.

Douglas, J. T., Goss, M. J., and Hill, D., 1980. Measurements of pore characteristics in a clay soil under ploughing and direct drilling, including the use of a radioactive tracer ($^{144}$Cs) technique. *Soil Tillage Res.,* 1:11–18.

Edwards, C. A. and Lofty, J. R., 1977. *Biology of Earthworms,* 2nd ed., Chapman and Hall, London, p. 333.

Ehlers, W., 1975. Observations of earthworm channels and infiltration on tilled and untilled loess soil. *Soil Sci.,* 119:242–249.

Ehui, S. K. and Hertel, T. W., 1989. Deforestation and agricultural productivity in the Cote d'Ivoire. *Am. J. Agric. Econ.,* 71:703–711.

Ehui, S. K. and Spencer, D. S. C., 1990. Indices for Measuring the Sustainability and Economic Viability of Farming Systems, RCMP Research Monograph No. 3, Resource and Crop Management Program, International Institute of Tropical Agriculture, Ibadan, Nigeria.

Ehui, S. K. and Spencer, D. S. C., 1992. Measuring the sustainability and economic viability of tropical farming systems: a model from sub-Saharan Africa. *Agric. Econ.,* 9:279–296.

Ehui, S. K., Hertel, T. W., and Preckel, P. V., 1990. Forest resource depletion, soil dynamics, and agricultural productivity in the tropics. *J. Environ. Econ. Manage.,* 18:136–154.

Evans, R., 1980. Mechanics of water erosion and their spatial and temporal controls: an empirical viewpoint, in *Soil Erosion,* Kirkby, M. H. and Morgan, R. P. C., Eds., John Wiley & Sons, Chichester.

Fragoso, C., Barois, I., Gonzalez, C., Arteaga, C., and Patron, J. C., 1993. Relationship between earthworms and soil organic matter levels in natural and managed ecosystems in the Mexican tropics, in *Soil Organic Matter Dynamics and Sustainability of Tropical Agriculture,* Mulongoy, K. and Merckx, R., Eds., John Wiley & Sons, Chichester, pp. 231–240.

Franzen, H., 1986. Physikalische Eigenschaften und Ertragsleistung eines Alfisols in Süd-West-Nigeria in Abhängigkeit von Bodenbearbeitung und Mulchbedeckung, Ph.D. thesis, Fachbereich Agrarwissenschaften, Georg August Universität, Göttingen, p. 136.

Getahun, A. and Jama, B., 1989. Alley cropping in the coastal area of Kenya, in *Alley Farming in the Humid and Subhumid Tropics,* Kang, B. T. and Reynolds, L., Eds., Proceedings of an international workshop held at Ibaden, Nigeria, March 10–14, 1986, IDRC, Ottawa, Canada, pp. 163–170.

Goldman, A., 1990. Diagnostic survey of fallow management systems in the forest zone of Nigeria, in Annual Report for 1988 of the Resource and Crop Management Program, International Institute of Tropical Agriculture, Ibadan, Nigeria, pp. 59–60.

Gorbenko, A. Y., Panikov, N. S., and Zbyagintsev, D. V., 1986. The effect of invertebrates on growth of soil microorganisms. *Mikrobiologia,* 55:515–521.

Hauser, S., 1990. Water and nutrient dynamics under alley cropping versus monocropping in the humid-subhumid transition zone. *Trans. 14th ICSS,* 6:204–209.

Hauser, S., 1993. Distribution and activity of earthworms and contribution to nutrient recycling in alley cropping. *Biol. Fertil. Soils,* 15:16–20.

Hauser, S., 1994. Soil and organic matter turnover by earthworms in cropping systems of the humid-subhumid tropics. *Trans. 15th ICSS,* 4b:100–101.

Hauser, S. and Kang, B. T., 1993. Nutrient dynamics, maize yield and soil organic matter in alley cropping with *Leucaena leucocephala,* in *Soil Organic Matter Dynamics and Sustainability of Tropical Agriculture,* Mulongoy, K. and Merckx, R., Eds., John Wiley & Sons, Chichester, pp. 215–222.

Hulugalle, N. R. and Ndi, J. N., 1993. Effects of no-tillage and alley cropping on soil properties and crop yields in a Typic Kandiudult of southern Cameroon. *Agroforestry Systems,* 22:207–220.

IBRD (International Bank for Reconstruction and Development), 1989. Sub-Saharan Africa: from Crisis to Sustainable Growth, World Bank, Washington, D.C.

Kang, B. T., Wilson, G. F., and Lawson, T. L., 1984. *Alley Cropping: A Stable Alternative to Shifting Cultivation,* International Institute of Tropical Agriculture, Ibadan, Nigeria.

Kang, B. T., Reynolds, L., and Atta-Krah, A. N., 1990. Alley farming. *Adv. Agron.,* 43:315–359.

Kollmannsperger, F., 1956. Lumbricidae of humid and arid regions and their effect on soil fertility. *Trans. Sixth ICSS, Rapp.,* C:293–297.

Kretzschmar, A. and Bruchou, C., 1991. Weight response to the soil water potential of the earthworm *Aporectodea longa. Biol. Fertil. Soils,* 12:209–212.

Kühne, R. F., 1993. Wasser — und Nährstoffhaushalt in Mais — Maniok — Anbausystemen mit und ohne Integration von Allekulturen ("Alley Cropping") in Süd — Benin. Hohenheimer Bodenkundliche Hefte 13, p. 244.

Lal, R., 1986. Conversion of tropical rainforest: agronomic potential and ecological consequences. *Adv. Agron.,* 39:173–264.

Lal, R., 1987. *Tropical Ecology and Physical Edaphology,* John Wiley & Sons, Chichester.

Lal, R., 1988. Effects of macrofauna on soil properties in tropical ecosystems. *Agric. Ecosystems Environ.,* 24:101–116.

Lal, R., 1991. Myths and scientific realities of agroforestry as a strategy for sustainable management for soils in the tropics. *Adv. Soil Sci.,* 5:91–137.

Lal, R. and Akinremi, O. O., 1983. Physical properties of earthworm casts and surface soils as influenced by management. *Soil Sci.,* 135:114–122.

Lal, R. and Cummings, D. J., 1979. Clearing a tropical forest. I. Effects on soil and microclimate. *Field Crops Res.,* 2:91–107.

Lal, R. and De Vleeschauwer, D., 1982. Influence of tillage methods and fertilizer application on chemical properties of worm castings in a tropical soil. *Soil Tillage Res.,* 2:37–52.

Lathwell, D. J., 1990. Legume green manures. Principles for management based on recent research, TropSoils Bulletin No. 90-01, Soil Management Collaborative Research Support Program, North Carolina State University, Raleigh, NC.

Lavelle, P., 1974. Les vers de terre de la savane de Lamto, in Analyse d'un Ecosysteme Tropical Humide: la Savane de Lamto, Bull. de Liaison des Chercheurs de Lamto. No. spec. 5:133–166.

Lavelle, P., 1978. Les Vers de Terre de la Savane de Lamto (Cote d'Ivoire): Peuplements, Populations et Fonctions dans l'Ecosystème, Thèse Doctorat, Paris, VI. Publ. Labo. Zool. E.N.S. 12, p. 301.

Lavelle, P., 1983. The soil fauna of tropical savannas. II. The earthworms, in *Tropical Savannas,* Boulière, F., Ed., Elsevier, Amsterdam, London, pp. 485–504.

Lavelle, P., 1994. Faunal activities and soil processes: adaptive strategies that determine ecosystem function. *Trans. 15th ICSS,* 1:189–220.

Lavelle, P. and Martin, A., 1992. Small-scale and large-scale effects of endogeic earthworms on soil organic matter dynamics in soils of the humid tropics. *Soil Biol. Biochem.,* 24:1491–1498.

Lavelle, P., Blanchart, E., Martin, A., Spain, A. V., and Martin, S., 1992. Impact of soil fauna on the properties of soils in the humid tropics, in Myths and Science of Soils of the Tropics, SSSA Special Publication no. 29.

Lee, K. E., 1985. *Earthworms: Their Ecology and Relationships with Soils and Land Use,* Academic Press, London, New York, p. 411.

Madge, D. S., 1965. Leaf fall and litter disappearance in a tropical forest. *Pedobiologia,* 5:273–288.

Madge, D. S., 1969. Field and laboratory studies on the activities of two species of tropical earthworms. *Pedobiologia,* 23:188–214.

Marinissen, J. C. Y. and Dexter, A. R., 1990. Mechanisms of stabilization of earthworm casts and artificial casts. *Biol. Fertil. Soils,* 9:163–167.

Martin, S. and Lavelle, P., 1992. A simulation model of vertical movements of an earthworm population (*Millsonia anomala* Omodeo, Megascolecidae) in an African savanna (Lamto, Ivory Coast). *Soil Biol. Biochem.,* 24:1419–1424.

Matlon, P. J. and Spencer, D. S. C., 1984. Increasing food production in sub-Saharan Africa: environmental problems and inadequate technological solutions. *Am. J. Agric. Econ.,* 66:671–676.

Moorman, F. R., Lal, R., and Juo, A. S. R., 1975. The soils of IITA IITA Technical Bulletin No. 3, International Institute of Tropical Agriculture, Ibadan, Nigeria, p. 48.

Mulongoy, K., 1990. Effect of wormcasts on successive maize crops grown on Alfisol, in Annual Report for 1988 of the Resource and Crop Management Program, International Institute of Tropical Agriculture, Ibadan, Nigeria, pp. 52–54.

Mulongoy, K. and Akobundu, I. O., 1985. Nitrogen uptake by maize in live mulch systems, in *Nitrogen Management in Farming Systems in the Humid and Subhumid Tropics,* Kang, B. T. and Van Der Heide, J., Eds., Joint publication: International Institute of Tropical Agriculture, Ibadan, Nigeria and Institute for Soil Fertility, Haren, The Netherlands, pp. 285–290.

Mulongoy, K. and Bedoret, A., 1989. Properties of worm casts and surface soils under various plant covers in the humid tropics. *Soil Biol. Biochem.,* 21:197–203.

Nye, P., 1955. Some soil-forming processes in the humid tropics. IV. The action of the soil fauna. *J. Soil Sci.,* 6:73–83.

OTA (Office of Technology Assessment), 1984. Technology to Sustain Tropical Forest Resources, OTA-TM-F-31, OTA US Congress, Washington, DC.

Parera, V., 1989. The role of *L. leucocephala* in farming systems in Nusa Tenggara Timur, Indonesia, in *Alley Farming in the Humid and Subhumid Tropics,* Kang, B. T. and Reynolds, L., Eds., Proceedings of an international workshop held at Ibadan, Nigeria, March 10–14, 1986, IDRC, Ottawa, Canada, pp. 143–153.

Pashanasi, B., Melendez, G., Szott, L., and Lavelle, P., 1992. Effect of inoculation with the endogeic earthworm *Pontoscolex corethrurus* (Glossoscolecidae) on N availability, soil microbial biomass and the growth of three tropical fruit tree seedlings in a pot. *Soil Biol. Biochem.,* 24:1655–1660.

Satchell, J. E., 1983. *Earthworm Ecology: From Darwin to Vermiculture,* Chapman and Hall, London, p. 495.

Sharpley, A. N. and Syers, J. K., 1976. Potential role of earthworm casts for the phosphorus enrichment of runoff waters. *Soil Biol. Biochem.,* 8:341–346.

Shaw, C. and Pawluk, S., 1986. Faecal microbiology of *Octolasion tyrtaeum, Aporrectodea turgida* and *Lumbricus terrestris* and its relation to the carbon budgets of three artificial soils. *Pedobiologia,* 29:377–389.

Spain, A. V., Lavelle, P., and Mariotti, A., 1992. Stimulation of plant growth by tropical earthworms. *Soil Biol. Biochem.,* 24:1629–1634.

Springett, J. A., Gray, R. A. J., and Reid, J. B., 1992. Effect of introducing earthworms into horticultural land previously denuded of earthworms. *Soil Biol. Biochem.,* 24:1615–1622.

Syers, J. K., Sharpley, A. N., and Keeney, D. R., 1979. Cycling of nitrogen by surface-casting earthworms in a pasture ecosystem. *Soil Biol. Biochem.,* 11:181–185.

Tiwari, S. C. and Mishra, R. R., 1993. Fungal abundance and diversity in earthworm casts and in uningested soil. *Biol. Fertil. Soils,* 16:131–134.

Tiwari, S. C., Tiwari, B. K., and Mishra, R. R., 1989. Microbial populations, enzyme activities and nitrogen-phosphorus-potassium enrichment in earthworm casts and in surrounding soil of a pineapple plantation. *Biol. Fertil. Soils,* 8:178–182.

Van der Meersch, M. K., 1992. Soil Fertility Aspects of Alley Cropping Systems in Relation to Sustainable Agriculture, Ph.D. thesis, No. 226, Fakulteit der Landbouwwetenschappen, Katholieke Universiteit te Leuven, Belgium, p. 179.

Watanabe, M., 1975. On amounts of cast production by the Megascolecial earthworm *Pheretima lupeinsis. Pedobiologia,* 15:20–28.

# CHAPTER 7

# Biological Management of Soil Fertility as a Component of Sustainable Agriculture: Perspectives and Prospects with Particular Reference to Tropical Regions

M. J. Swift

## INTRODUCTION

Hans Jenny described soil development as a function over time of the interaction of climate, parent material, topography, and biota (Jenny, 1941). While this paradigm was devised to account for the outcome of long-term processes, Jenny, in accord with other soil scientists, also recognized the importance of the biota to the more immediate properties of soil fertility. Despite this recognition of the significance of biological processes, soil biology as a discipline has historically played a relatively small role in the development of soil fertility management practices.

The role of science is to provide a predictive understanding of natural phenomena. Armed with this understanding, and within the limits of certainty that science can set, humans gain the potential to manage their physical and biological environment with insight and sensitivity. In ecology, with the study of the biological world at the scales of population, community, ecosystem, and landscape, the capacity for prediction has been limited in comparison with that in chemistry or physics or in other areas of biology such as physiology and genetics. Thus science-based soil management has until recently largely treated the soil as a physicochemical system. The biological components of soil management models are largely restricted to the physiological response of the plant to soil conditions. During the last two decades, however, priorities in soil management have shifted and led to the development of new approaches

1-56670-277-1/97/$0.00+$.50

that can be gathered together under the title of "biological management of soil fertility."

Biological management of soil fertility implies the harnessing of the biological resources of the ecosystem, particularly those of the soil itself, for the manipulation of soil fertility. We should be clear at the outset that this is not the same as organic farming (NRC, 1989). It does not eschew the use of inorganic inputs, but rather focuses on increasing the efficiency of their use by biological means. Emergence from the circumstances of nutritional deficiency and poverty that characterize the lives of many of the small-scale farmers of the world can only be achieved where there are sufficient resources to raise food productivity. With very few exceptions this will only be possible by the provision of external sources of some nutrient elements. Biological and physicochemical management should thus both be regarded as essential components of an integrated approach to soil fertility management.

This shift in emphasis has been summarized recently by the formulation of a new paradigm for soil fertility research, which asserts that we should, "rely on biological processes by adapting germplasm to adverse soil conditions, enhancing soil biological activity and optimizing nutrient cycling to minimize external inputs and maximize the efficiency of their use" (Sanchez, 1994).

Increased interest in a biological approach to soil fertility management is of course predicated on a significant maturing of the discipline that has taken place over the last two decades and that has been summarized in a range of reviews (Swift et al., 1979; Tinsley and Darbyshire, 1994; Fitter, 1985; Pankhurst et al., 1994; Woomer and Swift, 1994). It has also been driven by a variety of other causes. First, many farmers in developing countries, in contrast with those of the industrialized world, still have only limited access to inorganic fertilizers (FAO, 1993). This skewed distribution, and the high cost of fertilizer in most parts of the world, emphasizes the need for increasing the efficiency of their use, and many researchers see a combination of inorganic with organic sources of nutrient as the best route for this. Second, it is now apparent that many of the great gains in production made in the green revolution by use of high-yielding varieties with high inputs of inorganic fertilizer cannot be maintained indefinitely. Among the causes attributed to yield declines under long-term cultivation are changes in soil fertility associated with loss of organic matter and the accompanying decline in soil physical and chemical properties.

The third driving force for a more ecological approach to soil management has come from the sustainable development agenda in which central concern with the maintenance of yield is closely associated with desires to conserve natural resources, including a greater value accorded to maintenance of biodiversity. Forced to an extreme, sustainability may be seen as mutually incompatible with increased agricultural productivity. It has been interpreted by many agricultural research scientists, however, as signifying increased efficiency in resource use, including the need to utilize all available resources within economic limits that are realizable in the long term as well as profitable in the short term (Lynam and Herdt, 1989; Spencer and Swift, 1992).

Last, but by no means of least importance, the ecological approach to soil fertility management has been favored by the change in farming systems research to a more participatory and circumstance-related method of developing responses to farmer's constraints. This approach necessitates a more sensitive awareness of environment variation and its importance in regulating ecosystem function (Swift et al., 1994a).

This chapter will address the issue of biological management from the particular standpoint of research for increased productivity and sustainability in small-scale farming systems in the tropics, with particular reference to Africa. The urgent need for research to improve agricultural production in this region is seen from the way in which a variety of indicators in African agriculture all point in the same direction. Per capita food production is declining, associated with high rates of soil nutrient depletion, while fertilizer use is increasing at only a very slow rate and fertilizer use efficiency remains well below that found in the rest of the world (FAO, 1993; Stangel et al., 1994).

## BIOLOGICAL MANAGEMENT OF SOIL FERTILITY

### Soil Populations and Processes

The central paradigm for the biological management of soil fertility is "to utilize farmer's management practices to influence soil biological populations and processes in such a way as to achieve desirable effects on soil fertility" (Swift et al., 1994a). Biological populations and processes influence soil fertility in a variety of ways each of which can have an ameliorating effect on the main soil-based constraints to productivity:

1. Symbionts such as rhizobia and mycorrhiza increase the efficiency of nutrient acquisition by plants.
2. A wide range of fungi, bacteria, and animals participate in the processes of decomposition, mineralization, and nutrient immobilization and therefore influence the efficiency of nutrient cycles.
3. Soil organisms mediate both the synthesis and decomposition of soil organic matter (SOM) and therefore influence cation exchange capacity; the soil N, S, and P reserve; soil acidity and toxicity; and soil water-holding capacity.
4. The burrowing and particle transport activities of soil fauna, and soil particle aggregation by fungi and bacteria, influence soil structure and soil water regimes.

Specific examples of the ways in which these functions are performed and regulated are given in the other papers in this volume and in the reviews referred to earlier.

## Soil Management and Biological Processes

An understanding of the biological processes of soil is of no practical value except in the context of the regulatory influence of management practices. The range of management practices that a farmer can employ to regulate soil fertility is limited (Table 1). Most of these practices are not unique to biological management but common to farming practice the world over. They also have a history as long as that of agriculture itself. The most sophisticated expression of these practices is often to be found in so-called traditional agricultural systems such as shifting cultivation or valley-bottom rice production.

All of the practices listed in the first column of Table 1 influence soil biological populations and processes in a number of ways (column three). The most direct means of biological management are those associated with the use of biological inputs such as N-fixing bacteria, mycorrhiza, or soil fauna as a means of enhancing the endemic biological activities. These procedures are the focus of much contemporary research (e.g., see reviews by Giller and Wilson, 1991, and Pankhurst et al., 1994), but build on practices that farmers have utilized for many centuries. Direct management is also achieved by the use of organic matter inputs — in effect a means of selectively feeding the heterotrophic biological populations of soil — a practice of very ancient origin but at times eschewed in modern agriculture. Equally direct but usually unintentional effects are also achieved by the use of pesticides, which may kill particular groups of soil organisms that are involved in processes of significance to soil fertility. Management techniques such as tillage and fertilization also influence the activity of the biota indirectly by altering the physical and chemical environment of the soil.

Despite the fact that these relationships between management practices and soil biological activities have been known throughout most of the history of soil science, very little attempt has been made until recently to scientifically manage soil populations and processes. The capacity to do so rests, as prefaced in the introduction, in the ability to predict the outcome of the effects of management practice on soil biological activity and hence the impact on soil fertility. Successful biological management can only be said to be achieved when this interactive chain can be predictably followed through from input to outcome.

The success of biological management practices thus rests on two preconditions that must be satisfied: the availability of a management practice that is practically and economically acceptable to the farmer and the demonstration by the scientist that the practice leads to enhanced soil fertility. In the following sections these two aspects are considered (in reverse order) in relation to the use of organic inputs as a means of biological management.

**Table 1  Farmers' Management Practices for Influencing Soil Fertility Through Manipulation of Biological Processes**

| Management practice | Constraints to use | Biological processes influenced | Soil fertility effects |
|---|---|---|---|
| **Biological inputs** | Cost | | |
| Rhizobium inoculation | Availability of inoculum | N-fixation | Increased N-aquisition |
| Mycorrhiza inoculation | Environmental adaptation | Nutrient uptake by mycorrhiza | Increased efficiency of uptake of P and other nutrients |
| | | | Increased efficiency of $H_2O$ uptake |
| | | | Increased heavy metal tolerance |
| Soil fauna inoculation | Competition with fertilizer | Fauna burrowing | Soil structure/porosity |
| | Competition with indigenous biota | Decomposition | Stimulation of nutrient release |
| **Organic matter inputs** | | Decomposition | Increased short-term nutrient availability |
| Crop residues | Recycles nutrient only | SOM synthesis | Increased nutrient storage/exchange |
| Root residues | Inaccessibility to management | | Soil physical structure improved |
| Weed residues | Crop impact from competition | | Soil water regimes improved |
| Tree litters/prunings | Land set aside | | Acidity/toxicity diminished |
| Green manure | Land set aside | | Macropore formation improved (macrofauna) |
| Farmyard manure | Livestock and fodder availability | Soil fauna/microflora growth | Soil aggregation improved (microflora) |
| Household waste | — | | |
| Purchased organic input | Cost | | |
| Precomposting | — | | |
| | All: Labor availability | | |
| | Opportunity cost of other uses | | |
| Inorganic fertilizer inputs | Cost | Mycorrhiza inhibited | Direct transfer of nutrient to plant increased |
| | Availability | N fixation inhibited | Nutrient losses increased |
| | | Mineralization/immobilization balance changed | Acidification |

Table 1   Farmers' Management Practices for Influencing Soil Fertility Through Manipulation of Biological Processes (continued)

| Management practice | Constraints to use | Biological processes influenced | Soil fertility effects |
|---|---|---|---|
| Tillage | | | |
| Conservation | Soilborne disease | (Intensive tillage)<br>Decomposition stimulated by OM incorporation | (Intensive tillage)<br>Short-term nutrient availability increased |
| Hand | Labor | SOM decay stimulated by aeration and particle size reduction | Root growth in tilled layer promoted |
| Mechanized | Cost | Faunal and microbial populations diminished | Nutrient losses increased<br>Long-term nutrient storage diminished |
| Pesticides | Cost<br>Environmental and health impact | Nontarget organism populations diminished or eradicated | Destabilization of nutrient cycles<br>Loss of soil structure |

## MANAGEMENT OF ORGANIC INPUTS

### Regulation of Nutrient Dynamics by Resource Quality Control

Biological management of soil fertility depends on the manipulation of organic inputs to the soil more than on any other practice. The basic scientific premise for this practice is that the organic matter supplies energy and nutrients to the soil biota to stimulate their activities and therefore promote soil fertility.

Organic inputs are processed by a complex web of soil organisms (Figure 1), but biological management is based on the assumption that the outcomes are relatively consistent. These outcomes are multiple and include the generation of inorganic nutrients such as $NH_4$, $NO_3$, $PO_4$, and $SO_4$ from organic compounds of these elements; the synthesis of soil organic matter (SOM) from precursor compounds in the organic inputs; and modification of the physical structure of the soil as a result of stimulation of biological activities (Table 1, line 2). Quantitative variation in these outcomes occurs as result of the variety of potential organic inputs (Table 1), the diversity of organisms involved in the conversion processes, and the complexity of the interactions between them.

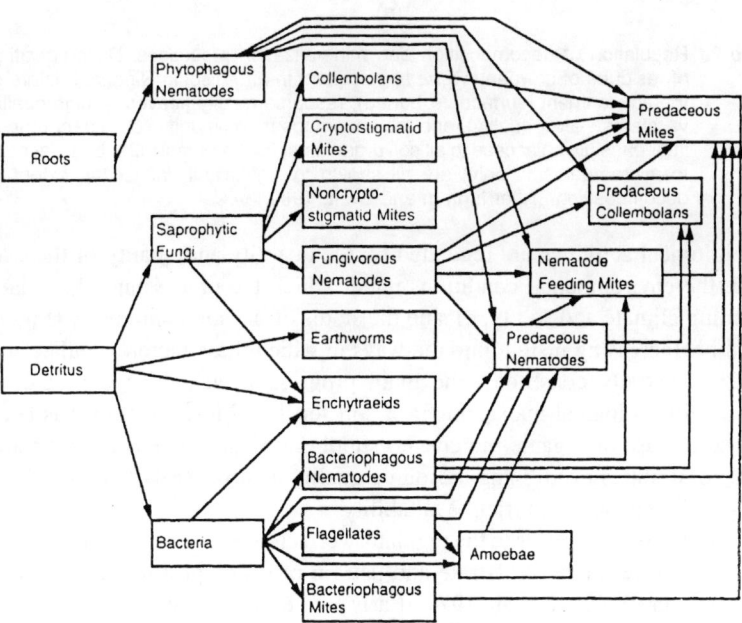

**Figure 1**    A crop residue (detritus)-based soil food web from a calcareous soil under arable cropping in The Netherlands. Under conventional tillage the bacteria-based food web is dominant; under reduced tillage the fungal and earthworm paths are more prominent (Verhoef and Brussaard, 1990). (From De Ruiter et al., 1993. *J. Appl. Ecol.,* 30:95–106. © 1993 with permission from Blackwell Science Ltd., Oxford.)

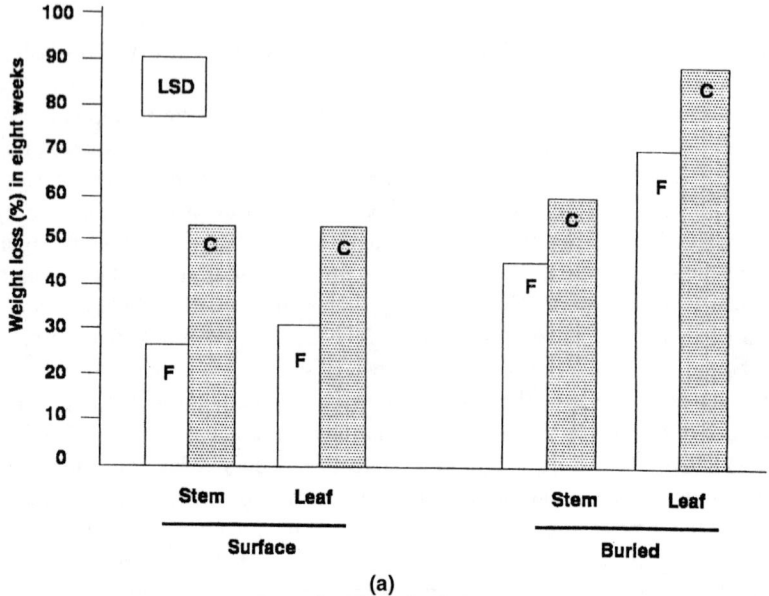

(a)

**Figure 2a** Regulation of decomposition and mineralization processes. Decomposition of residues of cowpea (*Vigna unguiculata*) in an Alfisol in Nigeria. Factors of the environment (surface vs. buried), resource quality (leaves — high quality vs. stems — low quality), and the decomposer community (*C* = coarse-mesh litter bags giving access to all soil organisms; *F* = fine-mesh litter bags, access to microorganisms only) are all shown to significantly affect the extent of decomposition. (After Ingram and Swift, 1989.)

The biological activities are regulated by the quantity and quality of the OM added, the environmental conditions under which the processing takes place (including climate and soil type), and the status of the soil community (Figure 2a and b). Increasing insight into the ways in which these factors regulate soil biological activity constitutes the main progress in achieving a predictive capacity for biological management of soil fertility. Most attention has been given to the use of organic mulches to supply nutrients such as nitrogen and to the possibilities of utilizing "resource quality control" (Swift, 1984; 1987) as a tool for managing nutrient availability.

Resource quality refers to the regulatory effect of the chemical composition of an organic resource on its rate and pattern of decomposition and nutrient release (Figure 2) (Swift et al., 1979). Early studies of decomposition processes demonstrated the importance of N concentration and C:N ratio as determinants of the N supplying capability of plant residues (Iritani and Arnold, 1960; Russell, 1961). Studies in natural ecosystems subsequently demonstrated that other indices such as the lignin concentration or the lignin:N ratio provided a better predictor of N release patterns (Melillo et al., 1982; Melillo and Aber, 1984). Berg and McClaugherty (1987) suggested that N is not released from

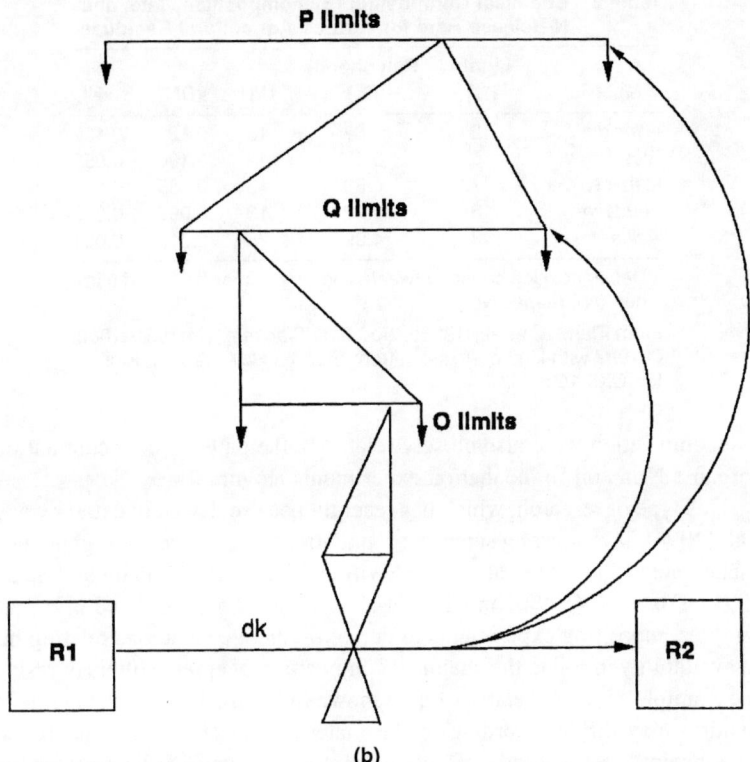

**(b)**

**Figure 2b** Hierarchical regulation of decomposition and nutrient release. $P$ = physico-chemical factors setting the broadest limits to the extent of decomposition (i.e., conversion of state R1 to state R2), these limits being successively fine-tuned by resource quality ($Q$) and the composition of the decomposer community ($O$). (After Swift, Heal, and Anderson, 1979.)

forest litters until decomposition of lignin commences, and others have also stressed the importance of the more recalcitrant components as regulators of decomposition rate and nutrient release (Feller, 1979; Gueye and Ganry, 1978). Polyphenols have also been implicated as regulators of N release in materials that, despite being nutrient-rich and low in lignin, are slow to release N (Vallis and Jones, 1973). It has been proposed that the polyphenols interact with N to form stable polymers that are resistant to breakdown and therefore delay N release (Martin and Haider, 1980; Stevenson, 1986).

These theories have been used to evaluate organic residues as sources and suppliers of nutrients within cropping systems. For example, Tian et al. (1992a, b) investigated the patterns of decomposition and nutrient release of a range of crop residues and agroforestry inputs under field conditions in a humid tropical environment. They found that decomposition and N-release were strongly correlated with N, lignin, and polyphenol concentrations (Table 2). The patterns of release of P, Ca, and Mg were similar to that of N. The rates

**Table 2   Chemical Composition, Decomposition Rate, and N-Release Rate for Various Agricultural Residues**

| Material | Lignin (%) | Polyphenol (%) | C/N | kDM[a] | kN[a] |
|---|---|---|---|---|---|
| Gliricidia | 12 | 1.62 | 13 | 0.127 | 0.152 |
| Rice straw | 5 | 0.55 | 42 | 0.106 | 0.052 |
| Maize stover | 7 | 0.56 | 43 | 0.085 | 0.077 |
| Leucaena | 13 | 5.02 | 13 | 0.062 | 0.055 |
| Acioa | 48 | 4.09 | 28 | 0.010 | 0.001 |

[a] Decomposition constant (week$^{-1}$) for dry matter (DM) and nitrogen (N), respectively.

From Tian, G. et al., 1992b. *Soil Biol. Biochem.*, 24:1051–1060. © 1992 with kind permission from Elsevier Science Ltd., Kidlington OX5 1GB, UK.

of decomposition were also closely related to the patterns of accumulation of inorganic N in soil in incubation experiments (Figure 3).

This type of research, which has recently been reviewed in detail by Myers et al. (1994), lays a strong scientific foundation for practices of organic matter management (Sanchez et al., 1989; Swift and Palm, 1995). Palm and Sanchez (1991), Fox et al. (1992), and Tian et al. (1992a) have all utilized information from decomposition experiments to derive predictive equations relating nutrient availability in soil to the chemical composition of applied litters or residues. One example of such a relationship is shown in Figure 4. While the appropriate equation may differ according to the materials chosen, it is clear that such relationships give a good prediction for the outcome of the application of organic residues to soil.

The same basic principles of decomposition regulation summarized in Figure 2b are also incorporated in a number of simulation models such as the Rothamsted Model (Jenkinson et al., 1987) and CENTURY (Parton et al., 1989), which have also been shown to have a high predictive power. These results give us confidence that we are developing tools that will enable organic inputs to be managed with a much higher degree of sensitivity and predictability in the future.

## Environmental Regulation of Organic Matter Management

Organic matter chemistry — resource quality — is not the only factor needed to make reliable predictions of the effects of organic matter inputs. In Figure 2b the physicochemical environment is pictured as exerting control over biological processes from a higher level in the hierarchy than resource quality. Much of this regulation such as that deriving from the climate lies outside the control of the farmer. There are practices, however, that can be manipulated so as to "environmentally tune" the influence of organic inputs on soil biological processes. These include the location and timing of the application of organic matter. It is well established, for instance, that incor-

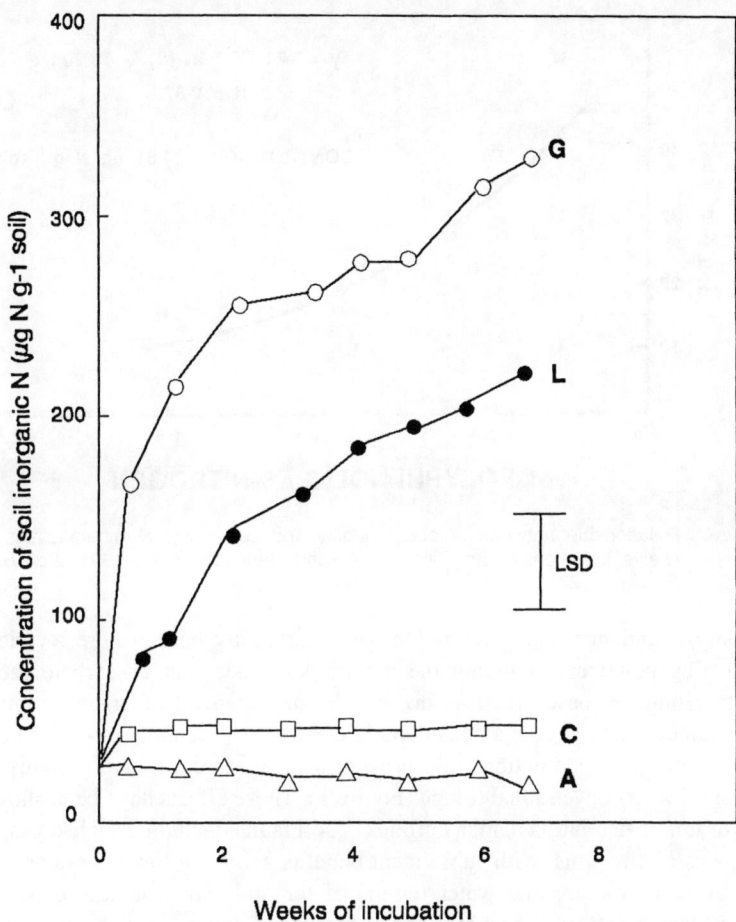

**Figure 3** Cumulative mineralization of inorganic nitrogen in soil mixed with residues of *Gliricidia sepium (G), Leucaena leucocephala (L)*, or *Acioa barterii (A)* compared with soil alone *(C)*. Compare these effects with the data for decomposition and mineralization rates in Table 2. (Redrawn from Tian, G. et al., 1992. *Soil Biol. Biochem.*, 24:1051–1060. © 1992 with kind permission from Elsevier Science Ltd., Kidlington OX5 1GB, UK.)

poration of residues in the soil, as opposed to surface mulching, can accelerate decomposition processes significantly and thus alter the dynamics of nutrient availability (Figure 5, a and b). The breaking up of the residues during tillage may be a factor in this effect, as well as the transfer of the organic matter into the more favorable environment below the soil surface.

The paradigm in Figure 2b also asserts that the pattern of decomposition is influenced by the nature of the soil community. Evidence for this as a significant effect is more tenuous and its relevance to soil management more controversial. It is well established that cultivation changes the composition,

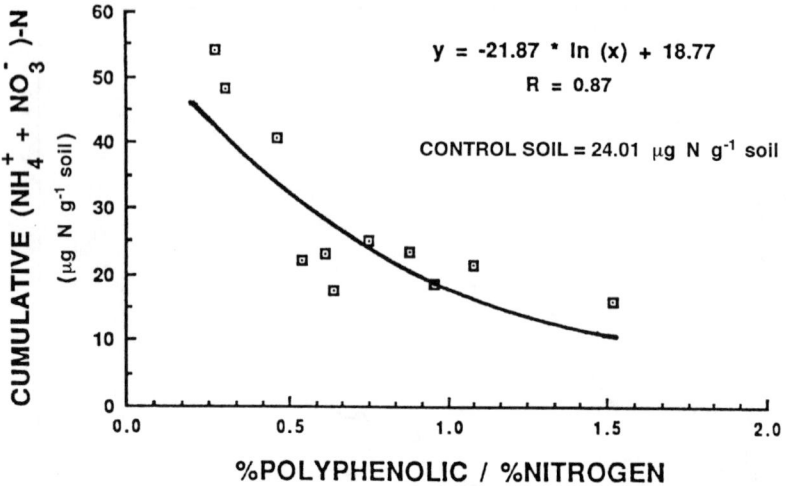

**Figure 4** Relationship between resource quality and cumulative N-mineralization in twelve legumes. (From Palm, C. A. and Sanchez, P. A., 1991. *Soil Biol. Biochem.*, 23:83–88.)

diversity, abundance, and activity of the soil community. Furthermore, a variety of farming practices, including the use of pesticides, can exacerbate these effects. Some of these practices involve the management of organic inputs. For instance, there is now a considerable body of data relating to differences in soil biological populations and processes in minimum-tillage systems as compared with conventional tillage (Figure 1). These effects have been shown to be dramatic in relation to many groups of soil fauna, including such keystone groups as earthworms, with subsequent benefits gained in low-till systems to the physical structure and water regime of the soil. The potential exists to utilize these practices to manipulate the biological community and thence to determine the outcome of decomposition processes. A major challenge is to determine pragmatically the extent to which these "fine-tuning" factors of microenvironment and community structure need to be taken into account when developing management systems.

## Organic-Inorganic Interactions

The use of organic inputs of different qualities to improve the efficiency of nutrient transfer to the crop will not be a practice of any significance if the total amount of nutrient available is insufficient to satisfy the needs for production. Significant input of N to replace that removed in harvest may be achieved in farming systems incorporating N-fixing species (Giller and Wilson, 1991). The amount of recyclable N is enhanced in tree-based systems such as traditional fallow rotations, savanna-based mixed farming systems, or modern

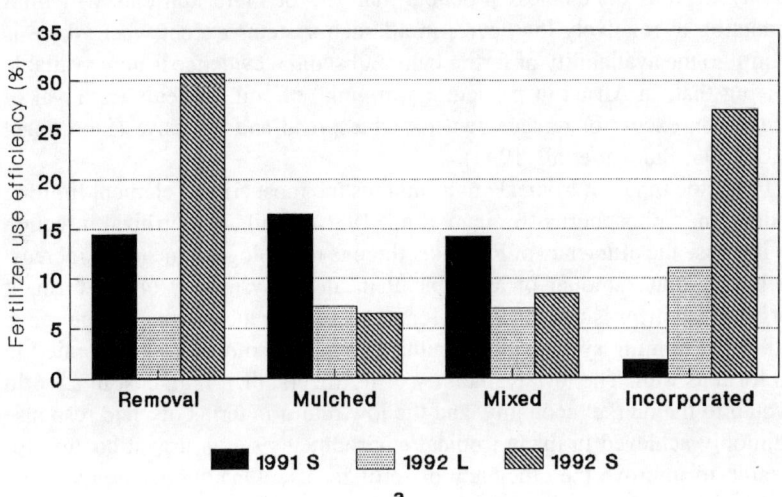

a

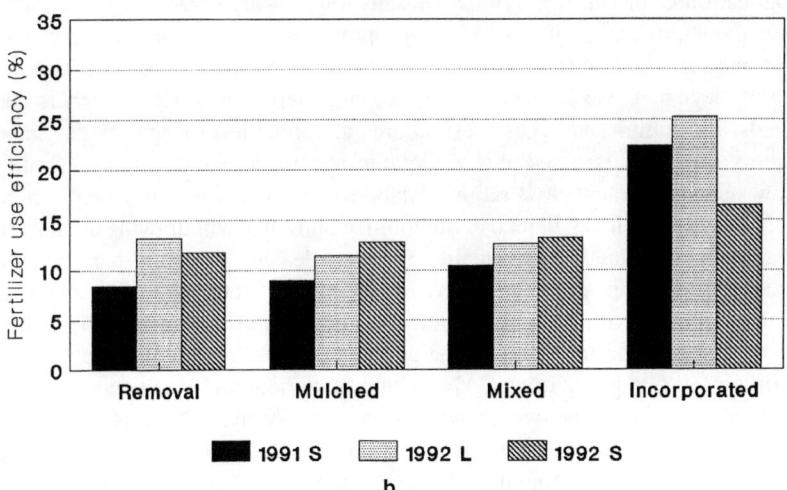

b

**Figure 5**   Fertilizer use efficiency (FUE) in maize cropping systems for three seasons at two sites in Kenya. FUE = (Grain-N for treatment — Grain-N in unfertilized control)/N added in fertilizer. The National Agricultural Research Laboratory (NARL) has an annual rainfall of 973 mm and is on a Nitisol; the National Dryland Farming Research Centre (NDFRC) has an annual rainfall of 673 mm and is on a Luvisol; L, long rain season; S, short rain season. Further explanation in text. (From unpublished results of S. Nandwa, with permission.)

agroforestry practices. These systems have the additional advantage of mobilizing other nutrients such as P and K by deep capture by tree roots from lower
soil layers. It is nonetheless probable that one or more nutrients will limit
production to relatively low levels in all such systems except where there is
no limit to the availability of fertile land. Substantial evidence is now available
showing that, in Africa in particular, "mining" of soil nutrients (removal of
elements in excess of replacement) is widespread and intensive (Stoorvogel
et al., 1993; Stangel et al., 1994).

In a wide range of tropical environments the most critical element limiting
production is phosphorus (Batiano et al., 1986). While mycorrhizal infection
can increase the efficiency of P uptake, there is no biological means of increasing the absolute amount of available P in an ecosystem. P in the form of
inorganic fertilizer is thus essential for increased productivity in a large range
of tropical farming systems. Other nutrients may also need to be supplied in
this form as well. The low availability of fertilizers, their high cost in a world
devoted to the market economy, and the low return in terms of yield response
commonly achieved in many tropical environments set an urgent context for
research to improve the efficiency of fertilizer use. One approach advocated
for this is management that incorporates both inorganic and organic inputs of
nutrients.

There is a long history of agronomic research on crop responses to the
application of mixtures as compared with single sources of nutrients. Judged
over the short term, that is within a cropping season, the results are variable.
In some cases mixtures of organic and inorganic fertilizers result in increased
yields beyond those achieved with inorganic fertilizer alone; in others the
yields are diminished. These results are interpretable in terms of the same
factors as those discussed above. Organic resources of low resource quality
(low nutrient content, low ratios of labile to recalcitrant components, high
polyphenol contents) will tend to immobilize nutrients, withdrawing them from
availability to plants. High-quality inputs, releasing nutrients early in the
process of decomposition, will serve to supplement inorganic fertilizers rather
than competing with them. The interaction of some of these effects described
above is illustrated in Figure 5. The diagram shows the estimated recovery of
fertilizer N (50 kg N ha$^{-1}$ as CAN) with and without added maize stover (4 t
ha$^{-1}$) at two sites over two growing seasons in Kenya. There is significant
interaction between the organic and inorganic input, but this effect is modified
by both macroenvironmental (between sites and between seasons) and
microenvironmental (surface vs. incorporated) effects. The organic-inorganic
interactions show both decreases of N-availability due to immobilization (e.g.,
NARL, short rains, incorporated) and increases due to mineralization (e.g.,
NDFRC, short rains, incorporated). The complexity of these responses clearly
illustrates the necessity for process-level controlled experimentation to dissect
out the effects of individual factors before predictive models can be derived.
Unfortunately, there has been very little process-level research on these inter-

actions so that management of such mixtures remains a largely empirical matter. This thus remains a high priority for research.

The most effective response of the crop to fertilization, whether from organic or inorganic sources, may come if the maximum demand for nutrients from the plant coincides in time and space with the availability of nutrients in the soil.

## Long-Term Effects

I have so far discussed the effectiveness of organic matter management in terms of short-term nutrient transfers. The major benefits may, however, be largely seen in the longer term. The strongest evidence for nutrient transfer from input to plant is obtained in studies using isotopically labeled material (Table 3). In the few examples available, the N use efficiency of plant residues by a first crop is low, in the order of 15% for legume residues and 5% for cereal straw residues, but there is much variation. This is broadly comparable with, or slightly lower than, the N use efficiency of inorganic fertilizer in the same environment. In the examples given there is, however, a consistently greater partitioning of N from the organic material to SOM than to the plant (Table 3). This suggests that residual effects are likely to be more important than those in immediate seasons. There is a substantial body of evidence to substantiate this point from long-term experiments in Africa and elsewhere (Swift et al., 1994b). Optimal development of integrated nutrient use systems will be achieved when it is possible to predict the outcome of a given strategy for use and management of nutrient resources (both organic and inorganic) over the long and short term. With regard to the use of organic sources of nutrients, the required information includes the amount or proportion of nutrients that will be released by the processes of decomposition, the time course of nutrient release, and the probable partitioning of the nutrients after release. Extending predictive capacity into these long-term, or residual, effects remains a major challenge. Significant advances have been made in recent times by the use of the simulation models referred to earlier. These models rely on

Table 3   Partitioning of N Added in Labeled Plant
Residues to the Subsequent Crop and to
Soil Organic Matter

| Authors | Crop (%) | Organic matter (%) |
|---|---|---|
| Ladd et al., 1981 | 11–17 | 72–78 |
| Janzen et al., 1990 | 14 | 21–40 |
| Ng Kee Kwong et al., 1987 | 11–14 | 73–84 |

From Myers, R. J. K. et al., 1994. *The Biological Management of Tropical Soil Fertility,* Woomer, P. L. and Swift, M. J., Eds., John Wiley & Sons, London, 81–117.

visualizing the SOM as divided into fractions of differing turnover times and with different significance to soil fertility (Jenkinson et al., 1987; Parton et al., 1989). The difficulty still remains of validating the models with components of organic matter that can be physically or chemically identified and measured in intact soil (Stevenson and Elliott, 1989; Swift et al., 1991).

## IMPLEMENTATION OF BIOLOGICAL APPROACHES TO SOIL FERTILITY MANAGEMENT

### The Farming Systems Context

This brief review provides evidence that, while there is still a good deal of research to be done, some of it in critical areas, manipulation of biological processes is a technically feasible approach to improvement of soil fertility management. Technical innovations are, however, of little use if they are not adoptable by farmers and if they do not meet, in the hand of the farmers, the essential tests of long-term productivity and sustainability. The history of agricultural development over the last several decades contains many examples of failures in adoption or sustenance of technologies that were judged scientifically sound (Lal, 1987). Examples of this within the area of resource management technology are the failures of water-management schemes in the inland valley zones of West Africa (Richards, 1985), minimum tillage practices in the same region (Lal, 1987), poor take-up of improved clearing methods (Bentley, 1986) and soil conservation methods, and the difficulties experienced more recently in promoting the spread of alley cropping (Dvorak, 1991).

The reasons for failure to adopt or persist with technology are as commonly due to social, cultural, or economic factors as to technical deficiencies. This necessitates an approach to technology development and transfer that integrates the criteria used by the farmer in deciding his or her actions, as well as those assumed in the scientists' models. Soil management is only one small component of a farmer's spectrum of activity. His or her knowledge and perceptions of soil-based constraints will be combined with a holistic appraisal of the potentials and limitations to production in a given year and then judged in relation to current production objectives and goals. Farmers' decisions are based on judgments between alternatives that relate to the whole range of environmental, biological, and economic information that is available. There are, however, commonly two major issues that most influence the decision: the farmers' production objectives and their assessment of the risk of any particular course of action. Thus new technologies, whether fertilizers or OM-promoting measures, that may maximize yields will only be accepted if they are consistent with both income-increasing and risk-avoidance objectives.

Technological change is thus more likely to succeed in terms of both adoption and of impact on production when it is adapted to farmers' circumstances rather than merely optimized in terms of scientists' criteria. This is

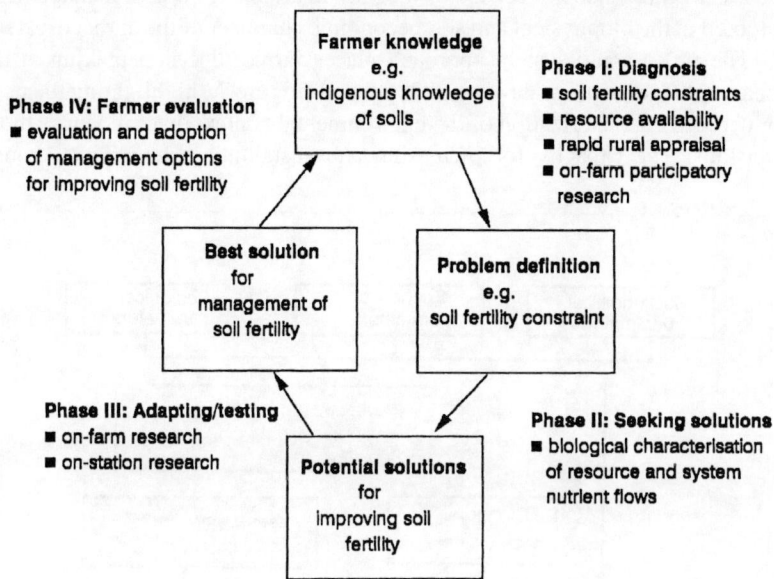

**Figure 6**   The "farmer-back-to-farmer" approach to agricultural problem identification and solution (Rhoades and Booth, 1982; Moran, 1987) as applied to soil fertility research. (From Swift, M. J., et al., 1994a. *The Biological Management of Tropical Soil Fertility,* Woomer, P. L. and Swift, M. J., Eds., John Wiley & Sons, London.)

most likely to be achieved when scientists and farmers work together throughout the research and development process. This approach is summarized in Figure 6, which illustrates stages in a research process that is initiated by joint diagnosis of soil fertility problems. This information, based both on farmers' perceptions and on biophysical and socioeconomic characterization of the farming system, can then be used to guide scientific investigation of the opportunities for improvement in soil management practices. Such research may require detailed process-level studies of the type described in the previous sections and should lead to recommendations for changed practices that are then tested and adapted by the farmer.

This interaction between process- and system-level research (Swift et al., 1994a) and between farmer-led and scientist-led investigation creates a continuous and integrated process of technology development and adoption that can be contrasted with the more familiar "technology transfer" approach where technological "improvements" are developed, largely by a process of empirical trial, in isolation from the farmer and in response to scientists' perceptions of generic productivity constraints, and presented as a package for adoption to the farmers in a given region. The process suggested here is not restricted to the concept of biological management of soil fertility — it is equally applicable to other aspects of agricultural development. It is, however, essential to the success of the concept, because of the way in which biological management, is rooted in the biophysical and socioeconomic character of the agroecosystem.

The process-and-system approach places farmer decision making at the center of the technology development process (Figure 7), highlighting the need for detailed characterization of the environmental context that influences those decisions. The capacity for increasing and sustaining productivity through

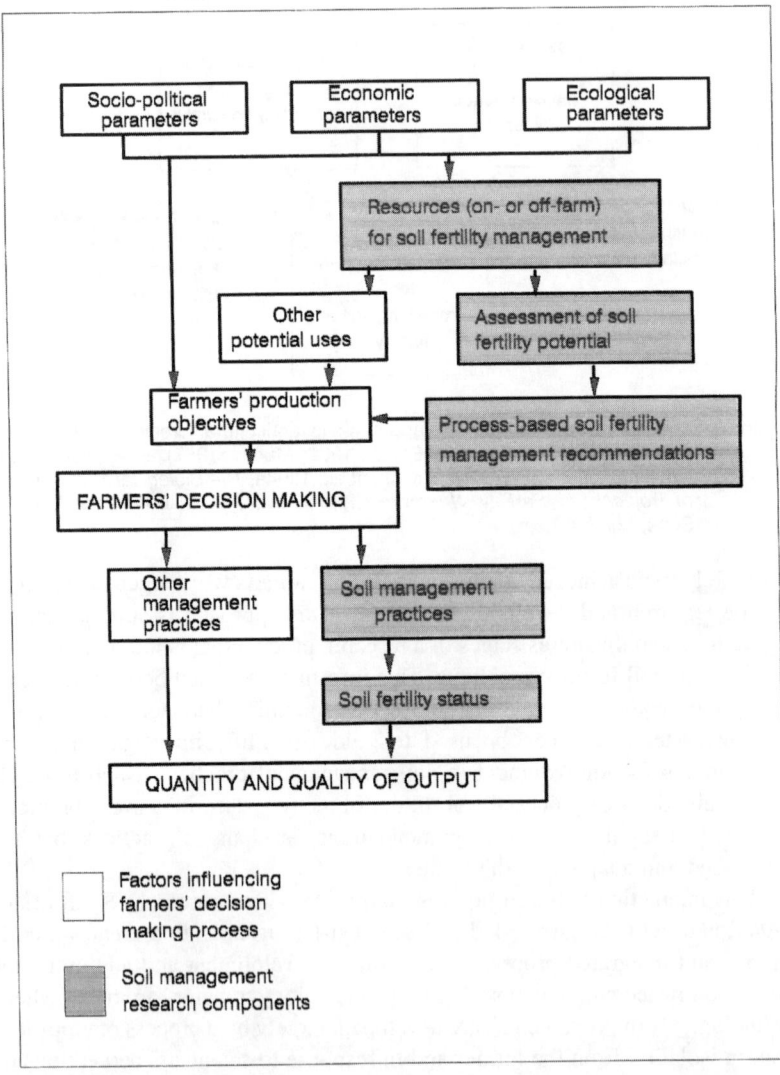

**Figure 7**    Integration of the research results for biological management of soil fertility with system level information on farmer's decision making. (From Swift, M. J., et al., 1994a. *The Biological Management of Tropical Soil Fertility,* Woomer, P. L. and Swift, M. J., Eds., John Wiley & Sons, London.)

improved soil fertility management is subject to a variety of limits of physicochemical, biological, economic, social, and cultural dimensions. The determination of these limits, the establishment of interactions — including hierarchical relationships, and the opportunities for modifying them is one of the major purposes of system characterization.

## CONCLUSIONS

The research required to harness the biological activities of soil for fertility management embraces the study of the populations of soil organisms, the processes they perform, and the factors that regulate them. From these process-level studies predictive models can be derived and used as a basis for soil management recommendations. This reductionist agenda is not sufficient, however, to achieve the goal of designing successful practices for soil fertility management. I have therefore dwelt on the circumstances of farmers and the context within which research for improved means of soil fertility management must take place. I have done this with the firm belief that it is on an understanding of these issues, as much as on the advances in soil biological understanding, that the future prospects for biological management of soil fertility and other aspects of sustainable agricultural development lie. Future research on the biological management of soil fertility must embrace this challenge if it is to achieve practical impact, as well as scientific illumination.

Smaling and coworkers (Stoorvogel et al., 1993) have analyzed nutrient deficiencies by using a simple input-output model. The results of such analysis at a variety of scales (Table 4) are stark, and the priorities these results indicate

Table 4    Average Nutrient Balances of N, P, and K (kg ha$^{-1}$ yr$^{-1}$) of Arable Land for Some Sub-Saharan African Countries

|  | N | | P | | K | |
|---|---|---|---|---|---|---|
| Country | 1982–84 | 2000 | 1982–84 | 2000 | 1982–84 | 2000 |
| Benin | −14 | −16 | −1 | −2 | −9 | −11 |
| Botswana | 0 | −2 | 1 | 0 | 0 | −2 |
| Cameroon | −20 | −21 | −2 | −2 | −12 | −13 |
| Ethiopia | −41 | −47 | −6 | −7 | −26 | −32 |
| Ghana | −30 | −35 | −3 | −4 | −17 | −20 |
| Kenya | −42 | −46 | −3 | −1 | −29 | −36 |
| Malawi | −68 | −67 | −10 | −10 | −44 | −48 |
| Mali | −8 | −11 | −1 | −2 | −7 | −10 |
| Nigeria | −34 | −37 | −4 | −4 | −24 | −31 |
| Rwanda | −54 | −60 | −9 | −11 | −47 | −61 |
| Senegal | −12 | −16 | −2 | −2 | −10 | −14 |
| Tanzania | −27 | −32 | −4 | −5 | −18 | −21 |
| Zimbabwe | −31 | −27 | −2 | 2 | −22 | −26 |

After Stoorvogel et al., 1993; Woomer, P. L. and Swift, M. J., Eds., 1994.

are fairly clear — the need to not only improve input availability but also the efficiency of nutrient use within systems. Biological approaches to soil fertility management can contribute to both these goals. The limits to production and sustainability are however far more complex than as simply indicated by the chemical inputs and outputs given in Table 4. The boundary conditions are themselves determined by the interaction of a variety of components of farmers' practices and farm history with the physicochemical, biological and socioeconomic environment. These factors not only determine current practices, but also influence the capacity and willingness of farmers to change practices and to adopt any new technology. Agroecosystem characterization and diagnosis, conducted at the farm system or village scale (Izac and Swift, 1994) can be iterated with scientists' process-level studies to provide an integrated approach to technology development (Figures 6 and 7).

The multifold increases in food production in Europe and North America, in the decades following the Second World War, and in Central America and Southeast Asia during the subsequent "green revolution" era, were achieved by a mixture of biological and physicochemical management. The biological management in these cases was of the crop plant. The sciences of genetics and physiology were brought to bear to produce crop plants with greatly enhanced photosynthetic potential. Management of soil and pests was largely implemented by physicochemical methods dependent on petrochemical-based industry. While very successful in terms of production, it is now realized that this approach brings a high environmental and economic cost.

A major challenge is thus to extend the reach of biology in an attempt to lower these costs. One route of this reach is down into the soil, with an increased emphasis on biological approaches to the management of soil fertility. As indicated in this review, this is fast becoming technically feasible as the science of soil biology hardens, but also requires a close integration with the sociocultural context of farming practice. The soil is part of a farmer's environment, as well as a resource to be exploited. These two features must be reconciled rather than seen as mutually incompatible.

## ACKNOWLEDGMENTS

The ideas of many friends and colleagues have contributed to this chapter among whom I should particularly like to acknowledge Cheryl Palm, Paul Woomer, Simon Carter, and Anne-Marie Izac and the comments of Ken Giller on an early version of the chapter. The mistakes and misinterpretations remain mine.

# REFERENCES

Batiano, A., Mughogho, S. K., and Mokwunye, A. U., 1986. Agronomic evaluation of phosphate fertilizers in tropical Africa, in *Management of Nitrogen and Phosphorus Fertilizers in Sub-Saharan Africa,* Mokwunye, A. U. and Vlek, P. L. G., Eds., Martinus Nijnoff, Dordrecht, The Netherlands, pp. 283–318.

Bentley, C. F., 1986. Disaster avoidance in land development projects, in *Land Clearing and Development in the Tropics,* Lal, R., Sanchez, P. A., and Cummings, Jr., R. W., Eds., A.A. Balkema, Rotterdam, pp. 3–16.

Berg, B. and McClaugherty, C., 1987. Nitrogen release from litter in relation to the disappearance of lignin. *Biogeochemistry,* 4:219–224.

De Ruiter, P. C., Moore, J. C., Zwart, K. B., Bouwman, L. A., Hassink, J., Bloem, J., De Vos, J. A., Marinissen, J. C. Y., Didden, W. A. M., Lebbink, G., and Brussaard, L., 1993. Simulation of nitrogen mineralization in the belowground food webs of two winter wheat fields. *J. Appl. Ecol.,* 30:95–106.

Dvorak, K. A., 1991. Methods of on-farm, diagnostic research on adoption potential of alley cropping. *Agroforestry Systems,* 15:167–182.

FAO, 1993. *Agriculture: Towards 2010,* FAO, Rome.

Feller, C., 1979. Une méthode de fractionnement granulométrique de la matière organique des sols, application aux sols tropicaux, à textures grossières, tres pauvres en humus, *Cahiers ORSTOM Séries Pédologie,* 17:339–346.

Fitter, A. H., Ed., 1985. *Ecological Interactions in Soil: Plants, Microbes and Animals,* Blackwell Scientific, Oxford.

Fox, R. H., Myers, R. J. K., and Vallis, I., 1990. The nitrogen mineralisation rate of legume residues in soil as influenced by their polyphenol, lignin and nitrogen contents. *Plant Soil,* 129:251–259.

Giller, K. E. and Wilson, K. J., 1991. *Nitrogen Fixation in Tropical Cropping Systems,* CAB International, Wallingford.

Gueye, F. and Ganry, F., 1978. *Etude de Compostage des Résidus de Recolte,* ISRA-CNRA, Bambey.

Ingram, J. A. and Swift, M. J., 1989. Sustainability of cereal-legume intercrops in relation to management of soil organic matter and nutrient cycling, in *Research Methods for Cereal/Legume Intercropping,* Waddington, S. R., Palmer, A. F., and Edje, O. T., Eds., CIMMYT, Mexico D.F.

Iritani, W. M. and Arnold, C. Y., 1960. Nitrogen release of vegetable crop residues during incubation as related to their chemical composition. *Soil Sci.,* 89:74–82.

Izac, A.-M. N. and Swift, M. J., 1994. An ecological-economic framework for developing sustainable agricultural technologies for sub-saharan Africa. *Ecol. Econ.,* 11:105–125.

Janzen, H. H., Bole, J. B., Biederbeck, V. O., and Slinkard, A. E., 1990. Fate of N applied as green manure or ammonium fertiliser to soil subsequently cropped with spring wheat at three sites in western Canada. *Can. J. Soil Sci.,* 70:313–323.

Jenkinson, D. S., Hart, P. B. S., Rayner, J. H., and Parry, L. C., 1987. Modelling the turnover of organic matter in long-term experiments at Rothamsted. *INTECOL Bull.,* 15:108.

Jenny, H., 1941. *Factors of Soil Formation,* McGraw-Hill, New York.

Ladd, J. N., Oades, J. M., and Amato, M., 1981. Distribution and recovery of nitrogen from legume residues decomposing in soils sown to wheat in the field. *Soil Biol. Biochem.,* 13:251–256.

Lal, R., 1987. Managing the soils of Sub-Saharan Africa. *Science,* 236:1069–1076.

Lynam, J. K. and Herdt, R. W., 1989. Sense and sustainability: sustainability as an objective in international agricultural research. *Agric. Econ.,* 3:381–398.

Martin, J. P. and Haider, K., 1980. Microbial degradation and stabilization of 14C-labeled lignins, phenols, and phenolic polymers in relation to soil humus formation, in *Lignin Biodegradation: Microbiology, Chemistry and Potential Applications,* Kirk, T. K., Higuchi, T., and Chang, H. M., Eds., Vol. 2, CRC Press, Boca Raton, FL.

Melillo, J. D. and Aber, J. D., 1984. Nutrient immobilization in decaying litter: an example of carbon-nutrient interactions, in *Trends in Ecological Research for the 1980's,* Cooley, J. H. and Golley, F. B., Eds., Plenum, New York, pp. 193–215.

Melillo, J. M., Aber, J. D., and Muratore, J. F., 1982. Nitrogen and lignin control of hardwood leaf litter decomposition dynamics. *Ecology,* 63:621–626.

Moran, E. F., 1987. Socio-economic aspects of research on tropical soil biology and fertility. *Biol. Int.* (Special Issue), 13:53–64.

Myers, R. J. K., Palm, C. A., Cuevas, E., Gunatilleke, I. U. N., and Brossard, M., 1994. Synchronization of nutrient mineralization and plant nutrient demand, in *The Biological Management of Tropical Soil Fertility,* Woomer, P. L. and Swift, M. J., Eds., John Wiley & Sons, London, 81–116.

Ng Kee Kwong, K. F., Deville, J., Cavalot, P. C., and Rivière, V., 1987. Value of cane trash in nitrogen nutrition of sugarcane. *Plant Soil,* 102:79–83.

NRC, 1989. *Alternative Agriculture,* National Research Council, National Academy Press, Washington, DC.

Palm, C. A. and Sanchez, P. A., 1991. Nitrogen release from some tropical legumes as affected by lignin and polyphenol contents. *Soil Biol. Biochem.,* 23:83–88.

Pankhurst, C. E., Doube, B. M., Gupta, V. V. S. R., and Grace, P. R., Eds., 1994. *Soil Biota: Management in Sustainable Farming Systems,* CSIRO, Australia.

Parton, W. J., Sanford, R. L., Sanchez, P. A., and Stewart, J. W. B., 1989. Modelling soil organic matter dynamics in tropical soils, in *Dynamics of Soil Organic Matter in Tropical Ecosystems,* Coleman, D. C., Oades, J. M., and Uehara, G., Eds., NifTAL, University of Hawaii, Honolulu, Hawaii, pp. 153–171.

Rhoades, R. E. and Booth, R., 1982. Farmer-back-to-farmer: a model for generating acceptable agricultural technology. *Agric. Admin.,* 11:127–137.

Richards, P., 1985. *Indigenous Agricultural Revolution: Ecology and Food Production in West Africa,* Unwin Hyman, London.

Russell, E. J., 1961. *Soil Conditions and Plant Growth,* 9th ed., Longmans, London.

Sanchez, P. A., 1994. Tropical soil fertility research: towards the second paradigm. *Trans. 15th World Congr. Int. Soil Sci.,* 1:65–88.

Sanchez, P. A., Palm, C. A., Szott, L. T., Cuevas, E., and Lal, R., 1989. Organic input management in tropical agroecosystems, in *Dynamics of Soil Organic Matter in Tropical Ecosystems,* Coleman, D. C., Oades, J. M., and Uehara, G., Eds., NifTAL, University of Hawaii, Honolulu, Hawaii, pp. 125–152.

Spencer, D. S. C. and Swift, M. J., 1992. Sustainable agriculture: definition and measurement, in *Biological Nitrogen Fixation and Sustainability of Tropical Agriculture,* Mulongoy, K., Gueye, M., and Spencer, D. S. C., Eds., John Wiley & Sons, Chichester, pp. 15–24.

Stangel, P., Pieri, C., and Mokwunye, A. U., 1994. Maintaining nutrient status of soils: macronutrients, in *Soil Resilience and Sustainable Land Use,* Greenland, D. J. and Szabolcs, I., Eds., CAB International, Wallingford, pp. 171–198.

Stevenson, F. J., 1986. *Cycles of Soil Carbon, Nitrogen, Phosphorus, Sulfur, Micronutrients,* Wiley, New York.

Stevenson, F. J. and Elliott, E. T., 1989. Methodologies for assessing the quantity and quality of soil organic matter, in *Dynamics of Soil Organic Matter in Tropical Ecosystems,* Coleman, D. C., Oades, J. M., and Uehara, G., Eds., NifTAL, University of Hawaii, Honolulu, Hawaii, pp. 173–245.

Stoorvogel, J. J., Smaling, E. M. A., and Janssen, B. H., 1993. Calculating soil nutrient balances in Africa at different scales. I. Supra-national scale. *Fert. Res.,* 35:227–235.

Swift, M. J., Ed., 1984. Soil biological processes and tropical soil fertility: a proposal for a collaborative programme of research. *Biol. Int., (Special Issue),* 5:1–37.

Swift, M. J., Ed., 1987. Tropical soil biology and fertility (TSBF): interregional research planning workshop. *Biol. Int., (Special Issue),* 13:1–64.

Swift, M. J. and Palm, C. A., 1995. Evaluation of the potential contribution of organic sources of nutrients to crop growth, in *Integrated Plant Nutrition Systems,* Dudar, R. and Roy, R. N., Eds., FAO Fertilizer and Plant Nutrition Bull. 12, FAO, Rome.

Swift, M. J., Heal, O. W., and Anderson, J. M., 1979. *Decomposition in Terrestrial Ecosystems. Studies in Ecology 5,* Blackwell Scientific, Oxford.

Swift, M. J., Kang, B. T., Mulongoy, K., and Woomer, P. L., 1991. Organic matter management for sustainable soil fertility in tropical cropping systems, in *Evaluation for Sustainable Land Management in the Developing World,* Vol. 2, Dumanski, J., Pushparajah, E., Latham, M., and Myers, R. J. K., Eds., IBSRAM, Bangkok, pp. 307–326.

Swift, M. J., Bohren, L., Carter, S., Izac, A. M., and Woomer, P. L., 1994a. Biological management of tropical soils: integrating process research and farm practice, in *The Biological Management of Tropical Soil Fertility,* Woomer, P. L. and Swift, M. J., Eds., John Wiley & Sons, London.

Swift, M. J., Seward, P. D., Frost, P. G. H., Qureshi, J. M., and Muchena, F. N., 1994b. Long-term experiments in Africa: developing a database for sustainable land use under global change, in *Long-Term Experiments in Agricultural and Ecological Sciences,* Leigh, R. A. and Johnston, A. E., Eds., CAB International, Wallingford.

Tian, G., Kang, B. T., and Brussaard, L., 1992a. Effects of chemical composition on N, Ca and Mg release during incubation of leaves from selected agroforestry and fallow species. *Biogeochemistry,* 16:103–119.

Tian, G., Brussaard, L., and Kang, B. T., 1992b. Biological effects of plant residues with contrasting chemical compositions under humid tropical conditions — decomposition and nutrient release. *Soil Biol. Biochem.,* 24:1051–1060.

Tinsley, J. and Darbyshire, J. F., Eds., 1994. *Biological Processes and Soil Fertility,* Martinus Nijhoff/W. Junk, The Hague.

Vallis, I. and Jones, R. J., 1973. Net mineralisation of nitrogen in leaves and leaf litter of *Desmodium intortum* and *Phaseolus atropurpureus* mixed with soil. *Soil Biol. Biochem.,* 5:391–398.

Verhoef, H. A. and Brussaard, L., 1990. Decomposition and nitrogen mineralisation in natural and agroecosystems: the contribution of soil animals. *Biogeochemistry,* 11:175–211.

Woomer, P. L. and Swift, M. J., Eds., 1994. *The Biological Management of Tropical Soil Fertility,* John Wiley & Sons, London.

# Index

## A

*Acaulospora mellea*, 93
Acetobacter spp., 101
Africa, soil-related studies in
  earthworms
    activity of
      methodological aspects of monitoring, 117
      nitrogen turnover, 127–130
      organic carbon concentrations and,
        127–130
      in permanently cropped fields, 124–126
      plant growth and, 130–131
      in relatively undisturbed environments,
        119–122
      slash-and-burn effects, 122–124
      spatial heterogeneity of, in alley cropping,
        126–127
      surface indexing, 115–116
    distribution, 117–119
    genera, 120
    materials and methods, 116–117
    species, 117–119
    surface casts
      description of, 119
      effect of radiation exposure, 123–124
      organic carbon levels, 121
      in permanently vs. newly cropped fields,
        123–124
  soil fertility
    biological management approaches
      impediments, 138–139
      implementation, 152–155
      management options, 140–142
      organic inputs, 141
      environmental regulation, 146–148
      long-term effects, 151–152
      nutrient dynamic regulation by resource
        quality control, 143–146
      organic-inorganic interactions, 148–151
      principles, 138
      processes, 140–142
      soil populations and, 139–140
Aggregates
  in conventional vs. integrated arable agricultural
    systems, 8–9

earthworm effects, 8
fungi and bacterial effects, 60
microaggregates
  description, 16
  in fine- and coarse-textured soils, 20–21
  role in physical protection of soils, 24
stabilization
  description, 60
  mechanisms, 63
Agroecosystems, low external-input
  mycorrhizal fungi interactions
    *Chinampas* agroecosystem, 94–95
    *Marceño* agroecosystem, 95–96
    other agroecosystems, 97–98
    *Stizolobium*-maize and squash rotation
      agroecosystem, 92–94
Alfisols, 125
Alley cropping
  description, 114, 126
  earthworm activity
    casting, 126–127
    description, 124
    spatial heterogeneity, 126–127
*Alternaria alternata*, 55
AM, see Arbuscular mycorrhiza
Amoebae, 47
Anaerobiosis, 11
*Apporrectodea caliginosa*, 50
Arable soils
  conventional agriculture
    detriments to soil, 91
    pesticide use, 2
    practices that effect mycorrhiza
      fertilizers, 102
      pesticides, 102–103
    soil ecosystem functioning in
      hydraulic conductivity, 9–10
      overview, 1–2
      porosity, 3, 5
      root–soil contact, 5, 7
      soil aggregates, 8–9
      topsoil, 7
      water retention in soil, 9–10
  fungal- and bacterial-based food webs
    distribution, 43
    vertical stratification, 43

161